Devon Estuaries

Devon Estuaries

A collection of essays on the estuaries of Devon
prepared by local amenity societies and conservation groups

Compiled and edited by Graham Wills

DEVON BOOKS

First published 1985 by Devon Books
Copyright © Devon County Council, 1985
ISBN 0-86114-763-4

British Library Cataloguing in Publication Data

Devon estuaries: a collection of essays on the estuaries of Devon
prepared by local amenity societies and conservation groups.
1. Estuaries—England—Devon
I. Wills, Graham
551.46'09   GC601

This book has been produced on behalf of the Amenities and
Countryside Committee of Devon County Council. It was compiled
and edited by Dr Graham Wills, a Senior Planning Officer in the
Amenities and Countryside Division of the County Property
Department. The Director of Property is Andrew Smy and the
Amenities and Countryside Officer is Peter Hunt.

DEVON BOOKS

Publishers to the Devon County Council

Devon Books is a division of A. Wheaton & Co. Ltd who represent
a consortium of three companies:

Design, Editorial & Publicity
Production & Manufacturing
A. Wheaton & Company Ltd, Hennock Road, Exeter, EX2 8RP
Tel: 0392-74121 Telex: 42749 (WHEATN G)
(A. Wheaton & Co. Ltd is a division of Pergamon Press)

Sales & Distribution
Town & Country Books, 24 Priory Avenue, Kingskerswell,
Newton Abbot, TQ12 5AQ Tel: 08047-2690

CONTENTS

ACKNOWLEDGEMENTS

The editor and publishers wish to thank the following for providing photographs reproduced in this book.
Bob Coe: p. 8
Devon County Council: pp. 4, 13, 15, 17, 23, 37, 38, 41 (foot), 44, 49, 58, 60, 61, 76, 83
C. Johnson: p. 74
Focus Photography: p. 3
The National Trust: pp. 35, 42
M. Parkinson: p. viii
P. C. J. Primmer: p. ix
R.A.F. Chivenor: p. 80
R.S.P.B.: p. 18 (foot)
Peter Thomas: pp. 7, 14, 18 (top, middle), 19, 20, 47, 57, 63
M. Vincer: p. 43
Graham Ward: cover
West Country Tourist Board: pp. 27, 34, 50, 51
R.E.L. Winjate: pp. 26, 29, 31, 32
Other photographs are by the editor.

Maps are reproduced from the Ordnance Survey 1:50 000 maps with the permission of the Controller of Her Majesty's Stationery Office, Crown copyright reserved.

Cover: *Aerial view of the Exe estuary*

Title page: *Dart estuary upstream of the Higher Ferry*

PREFACE

Devon is fortunate in having ten estuaries of its own and an eleventh that it shares with Cornwall. Each is different, but all are places of remarkable beauty and interest. Some have sheltered, secret places, the steep banks of their creeks and channels clothed in dense woodland. Others are spacious with vast mud-flats, sandbanks and salt-marshes stretching almost as far as the eye can see. For thousands of years these fertile grounds have supported a rich and varied community of plants and animals and have afforded a winter haven for migratory birds. They are rich in history. Generations of fishermen, explorers, colonists, sailors and adventurers have set sail from our estuary shores or have sought shelter and supplies there. For centuries they were routes for travel and trade in a county whose topography has impeded communications. Today they still have a major role, not least as a venue for boating and sailing holidays and all provide vital refuges for wildlife. Who better to describe the estuaries than the people who know and love them best – members of the amenity societies and conservation bodies who work, for the benefit of all of us, to safeguard the character and natural beauty of these areas from the pressures that threaten them. Ten such bodies and one private estate were invited to submit articles for this book. Grateful thanks are extended to them all, and to all who have contributed to this publication.

Axe estuary from Seaton Bridge

Otter estuary (right)

1

Axe Estuary

The Axe estuary is the most easterly in Devon. It is comparatively small, running only two and a half miles from the river mouth northwards to the present limit of tidal water just above Axe Bridge (1) near Colyford. From a viewing point by the Harbour Inn at Axmouth (2) one can see right across the river and the marshes beyond to the eastern edge of built-up Seaton. At its fullest, near Axmouth, the river itself is barely forty yards wide; at its narrow mouth, a pebble could be tossed from one side to the other.

It was not ever thus. In prehistoric and Roman times the river and estuary filled the whole valley, providing a safe, sheltered haven, with tidal waters reaching much further north than now. Ancient British trackways converged on what later became an important trade route. The Phoenicians may well have been the first to sail big ships right up the river. Later, major Roman roads – the Fosse Way and Icknield Way – added to the importance of the Axe haven. Stretches of paved Roman road have been found near to the entrance of Stedcombe Manor and in Axmouth; other evidence of significant Roman activity is still emerging today. Pulman, in T*he Book of the Axe*, writes: 'The mouth of the Axe was at that time a noble estuary extending up to Colyford, covering the whole of the area of what is now the marsh.' At low tide, huge areas of tidal mud-flats were exposed; what little salt-marsh there was developed in the up-river creeks. Recent excavations for a new flood bank across the marshes revealed the blue estuarine mud and the fossils character- istic of tidal mud-flats.

From the end of the last phase of the Ice Age, continuous geological changes occurred which fundamentally altered the relation- ship of sea-level and coastal land mass. The decisive factor, however, was the building up at the mouth of the river of a huge pebble and shingle bank. This was due to the peculiar west-to-east drift of the tides and currents of Lyme Bay. Slowly but inevitably, this bank began to close the river mouth. The Axe, in addition, suffered from landslips at Haven Cliff. By the end of the thirteenth century, the whole character of the estuary had changed and it was inaccessible to all but the smallest craft.

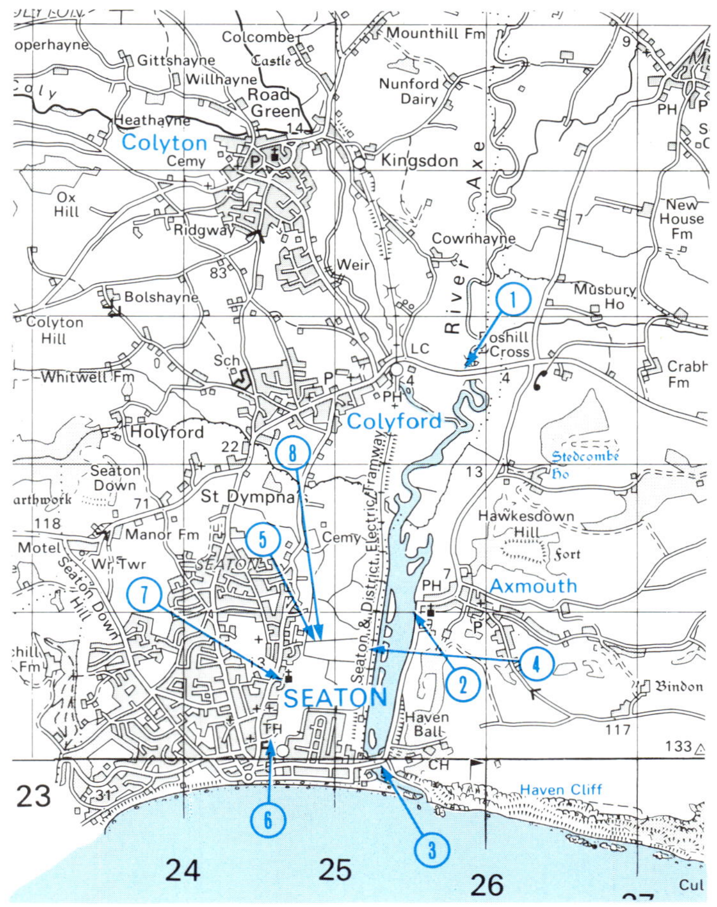

Motorway bridge crossing Exe estuary reed beds (above)

Aerial view of the Teign estuary (right)

Aerial view of the Axe estuary

Leland, antiquary to King Henry VIII, made a survey of England in 1533 – 43 and afterwards summarized his observations in his *Itinerary*. Of the Axe estuary he wrote: 'Ther hath beene a very notable Haven at Seton but now ther lyith betwen the 2 Pointes of the old Haven a mighty Rigge and Barre of pible Stones in the very Mouth of it and the Ryver of Ax is dryven to the very Est Point of the Haven, caullid Whit Clif, and ther at a very small Gut goith into the sea; and her cum in small fisher Boates for sucour. . . The Town of Seton is now but a meane Thing, inhabited with Fischar men; it hath bene far larger when the Haven was good.'

This just about sums up the decline of the haven. Even so, Leland wrote of Axe Bridge near Colyford, 'this Bridge servith not to pass over at High Tides; otherwise it doth'. This shows that the estuary still stretched further inland than it does now. And he had earlier written: 'I passid from Seton at Ebbe over the Salt Marshes and the Ryver of Axe to Axmouth'. The river is not fordable there now. At the

Seaton end, serious weaknesses have recently been identified in the structure of the old concrete bridge **(3)** and this historic monument is due to be replaced by a new road bridge. The original one, said to be the oldest concrete bridge still in regular use in Britain, may survive for pedestrian use.

The build-up of the shingle bar did more than contribute decisively to the demise of the old Axe haven. Behind it, salt-marsh developed where previously only tide-washed mud-flats existed. New, more solid land was built up. A whole new range of flowering plants was able to colonize the land, including glasswort, sea purslane, sea thrift, sea aster and sea lavender.

The salt-marshes were used for grazing, despite flooding by spring tides, and the temptation grew to build sea-walls and ditches to protect the pasture. In the 1660s the Lord of the Manor of Seaton, John Willoughby, built a 'reclaiming bank' **(4)** which stretched from the northern boundary of the marshes right down to the shingle bar, cutting off the southern marshes from tidal water. Sluices in the bank let fresh water out at low tide and kept the salt water out at high tide. The old salt-marshes became highly esteemed freshwater grazing meadows.

Another development in the early eighteenth century was the revival of salt-making. Special sluices had to be built to channel salt water into what was 'reclaimed land'. Traces of eighteenth-century salt-pans can still be identified and part of the marsh was called 'Salt Plot' in the Seaton Tithe Map of 1840.

In 1868 the building of the railway embankment – using the line of the original Willoughby 'reclaiming bank' but extending much further north – finally established the pattern of the estuary and marshes which we have inherited. Little true salt-marsh survived.

The Axe estuary today

What will the visitor find today? The estuary and marshes must be thought of as an entity, each area being complementary to the other. For convenience, however, they can, like Caesar's Gaul, be divided into three parts: the upstream area; the lower tidal reaches; and the marshes.

The upstream area

This area lies between a new flood bank **(5)** and the new Axe Bridge near Colyford. It includes, on both sides of the river, mud-flats and marshes, both brackish and salt, reed-beds, short stretches of the River Coly and of the Stafford Brook, extensive grazing meadows and even a football pitch. There is no public access, though extensive views can be enjoyed from several vantage points on the roadside on the east and by passengers on the trams. It includes some of

Axmouth harbour, near the mouth of the estuary

Axe Bridge, Seaton

the most attractive resting and feeding grounds for wintering birds. The land is privately owned, some grazing land being leased to local farmers.

The lower tidal reaches

This area, from immediately above the Harbour Inn, Axmouth, down to the Seaton Bridge and the river mouth at Haven Cliff, includes fragmented remnants of salt-marsh and is the focal area for birds throughout the year. During autumn and winter, the tidal waters, the low-water mud-flats and the adjoining banks, meadows and marsh-land provide a haven and feeding ground for a remarkably wide range of species, from geese to grebes and from wigeon to waders. Many species come in as migrants to breed in summer; others, such as the osprey and the avocet, may pass through in spring and autumn. There are many interesting resident and breeding birds as well, including the kingfisher, shelduck and heron.

The whole area can at all times be observed from the roadside between the Seaton Bridge and the Harbour Inn at Axmouth where there is also an open grassed area off the road, with seats and a Heritage Coast information board. If the roadside – with the possibility of bird-watching from the car – can be compared to a long static hide, the tramway (see below) provides a mobile hide for passengers. No special facilities are needed for handicapped or elderly persons to be able to enjoy the views and watch the ever-changing bird scene.

The marshes

This area comprises the marshland south of the new flood bank and west of the old railway embankment (now the tramway). It is the main expanse of open marshland, mostly grazed and criss-crossed by drainage ditches. It provides resting and feeding ground for the winter birds, particularly at high tide when they are forced off the mud-flats. The area is thus directly complementary to the tidal reaches and equally important.

On the western side there are wild flowers, including two species of orchid, and vegetation, bramble and a few shrubs which attract smaller nesting birds (such as stonechat), as well as butterflies, moths, and other insects of interest. Kingfishers are often to be seen along the ditches; kestrels hunt for small rodents.

A public footpath runs northwards from the Harbour Road car park **(6)** to St Gregory's Church **(7)** and is due to be extended to the Hillymead estate. It provides a good view of the open marshland. There are tentative plans, incorporated in the Seaton Local Plan proposed by East Devon District Council, for a small local nature reserve in this area. Meanwhile, a family picnic area and some other amenity developments of a modest nature have been proposed on a small site just north of the public car park.

The borrow-pit lagoon

In addition, there is a special small conservation area at the west end of a new flood bank, the construction of which left a sizeable flooded borrow-pit **(8)**. Here the Axe Vale and District Conservation Society has been granted a licence by South West Water for the protection and maintenance of the borrow-pit lagoon, with its little island and surrounding area, as a refuge and breeding, resting and feeding place for birds. Some plants and shrubs have already been established on the island. There is a need at all seasons for an absolute minimum of noise and disturbance, so there is no public access to the area.

The tramway

A unique feature of the estuary is the existence of an electric tramway which runs through the middle of the area, along what used to be the railway embankment. From Easter to September old-fashioned-looking but efficient and comfortable double-deckers, open on top, trundle up and down from Seaton to Colyton, laden with holiday-makers. In other months the public service is limited.

The very thought of such a mechanized invasion of marshland will no doubt horrify many readers unfamiliar with the area. And it is probably true that, if the tramway did not already exist, any planning application for one as a new feature would today arouse cries of alarm and distress from all manner of conservation interests, local, regional and even perhaps national.

In practice, the tramway is a positive asset. In summer it provides innocent pleasure to thousands of visitors, while doing no harm to wildlife. Its very existence helps to counter any charges that conservationists and bird lovers selfishly want to keep every-one else away from 'their' marshes. More positively, in autumn and winter months, the best for seeing wildfowl and waders, the trams offer organized parties splendid 'mobile hides' from which to observe the teeming bird life of the estuary and marshes. A typical winter list could include numerous shelduck, wigeon, teal, oyster-catcher, lapwing, dunlin, redshank, curlew, heron, ringed plover and, singly or in smaller groups, little grebe, cormorant, grey plover, spotted redshank, greenshank, common and green sandpipers, bar-tailed godwit and snipe, not to mention various gulls and an Axe speciality, the kingfisher. Trams can be privately chartered at a very reasonable charge per head by groups or societies. (Inquiries should be addressed to the Seaton and District Electric Tramway Company, Riverside Depot, Station Road, Seaton, Telephone Seaton 21702.)

The trams potter up and down the marshes, stopping frequently for passengers to point out birds to each other or just to enjoy the varied scenery. The trams get through to parts nothing else can

Curlew

reach! Moreover, such expeditions do not disturb the birds, which in any case are constantly on the move as tidal and other conditions change. Passengers should not, of course, create too much noise or disport themselves too energetically at stops.

An impressive demonstration of all this was organized by the Axe Vale and District Conservation Society, when a three-tram trip was undertaken by ninety members in mid-February 1984. On a clear, cold but sunny morning more than forty species of birds were identified. High tide had pushed most of the wildfowl and waders off the river banks and mud-flats on to the marshes, flooded meadows and dykes. As the tide went down and favourite feeding grounds reappeared, there was continual movement and excitement among the birds which gave watchers the feeling of being right in the middle of it all – as indeed they were. No rarities were recorded – a recently reported laughing gull had departed – but a very respectable list included several birds of special interest such as green sandpiper, bar-tailed godwit, little gull, spotted redshank and, of course, a kingfisher flashing past. By happy coincidence, a group of members of the East Devon Branch of the Devon Bird-watching and Preservation Society was patrolling the roadside across the river. Caught in this cross-fire, if that is the word, how could any bird have escaped identification?!

Bird-watching

An indication has already been given of the estuary's rich variety of birdlife. For such a small area it does seem to have more than its fair share of resident, wintering and migratory birds. Some rarer species have been recorded in recent years: smew (January 1979), avocet (April 1979), osprey (September 1981), white-fronted and greylag geese (January 1982), flamingo (August 1982), little egret (August 1983), laughing gull (January 1984). Dedicated bird-watchers will probably find a winter visit most rewarding but there is a lot to be seen at all seasons. Some of the birds can of course be seen out of their main season, for example early or late arrivals on migration. Exceptional weather affects the pattern, and some birds may well not appear every year, if ever again. New species are equally likely to appear. The state of the tide is important: local tables should be consulted.

Future developments

The Axe estuary and marshes are subject to all-too-familiar pressures. There are the inevitable farmers who 'need' to drain just those few extra acres for grazing. Local councils and other bodies with commercial and other interests yearn to develop this or that area for what in planners' jargon are called 'public recreation and amenity' purposes. (In fact, the estuary, left as it is, could have considerable tourist potential, particularly for bird-watchers in winter.) Local bird-watchers and conservationists try to be reasonable and constructive while resolutely resisting anything which would significantly alter the basic character of the estuary and marshes. Indeed, the Axe Vale and District Conservation Society, with its six hundred members behind it, has put forward positive ideas of its own and, in the case of the borrow-pit lagoon mentioned above, actually achieved something of practical value.

The basic point is that the estuary and marshes, *as they are*, undeveloped and unimproved, with the tramway and all, already provide a unique public amenity which, if cherished and protected by the local community itself, can continue to be a valuable local asset, an attraction for visitors, a rich haven for birds, and truly a 'thing of beauty and a joy for ever'.

Members of the Axe Vale and District Conservation Society bird-watching on the estuary

This chapter was written by Mr Philip R. Noakes on behalf of the Axe Vale and District Conservation Society of which he is Vice-Chairman. The author gratefully acknowledges the assistance of Mrs Margaret Parkinson, for advice on ecology, geology and history (readers wishing to study the subject further could do no better than read Mrs Parkinson's full exposition due to be published in the 1985 Transactions of the Devonshire Association for the Advancement of Science, Literature and Art); Mr Geoffrey Chapman, for help on local history and birds; Mr A. R. Hodge, for help on birds; and many other local residents who have provided material and advice.

The Axe Vale and District Conservation Society was founded in 1973. Its main aim is to protect and to promote public interest in the wildlife and the natural amenities and beauty of the Axe Valley. It has more than 600 members and publishes a biannual Newsletter. Enquiries about membership and current activities could conveniently be addressed to the Vice-Chairman, Mr P. R. Noakes, O.B.E., Little St Mary's, Uplyme, Lyme Regis, Dorset DT7 3XH.

2

Otter Estuary

Between the eastern limit of the Pebble Beds which are exposed in the cliffs at Budleigh Salterton, and a ridge of Otter Sandstone which forms the cliffs of Otterhead (1) lies the Otter estuary. It is a compact estuary with the river running into the sea on the eastern side close under the red cliffs of Otterhead. A curving line of rocks, known as Otterton Ledge, runs from the base of the cliffs seawards on the east side of the river mouth. These rocks form a small bay which helps to protect the eastern end of Salterton beach from erosion. A line of pine trees stands out on the top of the low cliffs which stretch northwards up the valley on the east side. At the foot of the trees many people tread the recently designated Coast Path which follows the line of the river to the first crossing place from the sea at White Bridge (2). This path behind the trees allows excellent viewing places from which the salt-marshes with the river, the cliffs, the rocks, the sea and headlands beyond may be seen. Newly ploughed fields on the slopes of the valley of the River Otter glow a deep red as the river has cut its way through the sandstone lowlands of the Devon Redlands to reach its present level in a shallow, mainly broad valley. Its more recent course through the alluvial soils of the once-wide estuary was man-made in the early nineteenth century. The parish boundary between Otterton and East Budleigh, established in the twelfth century as the middle of the river, now meanders through the lower valley and the estuary salt-marsh.

There is a footpath at the west side of the river valley which begins from Lime Kiln car park (3) at the seaward end of Granary Lane. If you follow this you will notice on the left wave-worn red sandstone cliffs which occur intermittently up the valley nearly to Pulhayes Farm in Lower Budleigh. The cliffs are now partly hidden by some fine oak trees with undergrowth of alder, hawthorn, elder, sycamore and wild rose. Nomadic hunters from the Palaeolithic Age (about 250 000 B.C.) may have sheltered in the sandstone cliffs on either side of the valley and hunted fish and wildfowl in the estuary, though the only evidence found so far of their possible presence is a crude hammer stone from Collever, east of Otterton. There is ample evidence in stone artefacts, made from local chert, of the

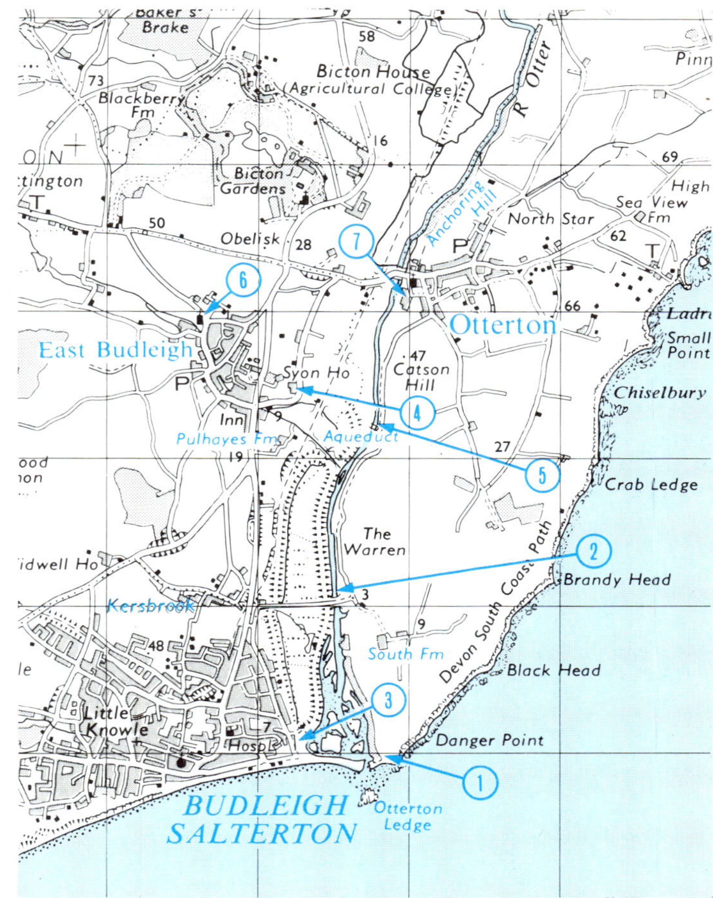

The Otter estuary, looking south towards Otterhead

presence of the later Mesolithic hunters (about 8500 B.C.). It is thought that they used fire as a tool to make clearings in the forest in which it was easier to hunt wild animals. The tools found indicate campsites where the kill was prepared and hides dressed. It was not until after 3000 B.C. and the invasion of Neolithic farmers from the Continent that the valley landscape began to include cultivated crops and penned animals. The settlers cleared land of trees and built earth banks to confine their domesticated animals. Over the centuries trade and travel developed and the estuary gained importance for additional reasons but its use for agriculture and fishing continues to the present day.

A drainage dyke runs to the right of the western path where once the river flowed. The first Ordnance Survey map, published in 1809, five centuries after the estuary began to be blocked by a pebble spit, still shows the extent of the old river mouth covering what is now the car park, cricket field and pastures. The river appears to have flowed along the west side of the valley at least as far as Thorn Mill **(4)** in Lower Budleigh. In summer, luxuriant riverbank flora remains in damp stretches to the left of the path and beside the dyke. This includes a fine stand of common reed mace (bulrush), reed, horsetail, gipsywort, meadowsweet, branched bur-reed, water pepper, yellow iris and great willow herb among many others.

The tidal River Otter above White Bridge

Mallard and moorhen frequent the dyke while reed warblers sing from the reeds. Near the modern road to South Farm the footpath crosses a stream flowing from Kersbrook to the west. The 1809 map depicts this area as a wide creek where the water washed the cliffs at high tides and flowed up to Kersbrook.

The modern footpath continues on the north side of the South Farm road with cliffs close on its left-hand side and a recently constructed pond on the right. Alder and white poplar shade the pond where moorhen busily feed. The rich riverbank flora continues, with the addition in this stretch of orange balsam. Grey wagtail and kingfisher may be seen near the dykes and, in winter, Canada geese sometimes graze in the fields with snipe hunting feverishly in the tussocks for insects. In the distance a line of poplar trees crosses the valley, the branches framing Anchoring Hill in a reach of the valley above Otterton. A short way beyond the poplars on the left, a side track joins the footpath. This track runs from East Budleigh Road and passes under the railway embankment. At about this point on the 1809 map, long before the railway was constructed, a track is shown coming to the river from Lower Budleigh. Could this area below Pulhayes Farm be the location of the port of Budleigh from which small ships once sailed to take part in the Hundred Years War and to carry wool as far as Spain? In 1347 one hundred and forty-one men of Budleigh are recorded as lost with three ships and twelve boats in an attack by French pirates as the fleet was returning from Spain. Was this the place from which it had sailed?

The present footpath runs in a broad curve across what may have once been water, though in 1809 it was mapped as marsh, to the river at a point below Clamour Footbridge (5) where an aqueduct, carrying Budleigh Brook, empties into the main river channel. Here the willow trees beside the main river attract many insects which are food for various birds. Tree creepers, goldcrests, and tits move rapidly as they hunt in the trees, while grey wagtails and common sandpiper search along the water's edge. Where the water of Budleigh Brook enters the Otter, water mint and monkey flower grow. At the river the footpath divides. A branch veers to the left leading to East Budleigh by way of Thorn Mill Farm. The main path continues beside the river to Otterton.

The port of Ottermouth, thriving until the early fourteenth century, was stated by Leland (writing in 1543) to be about a mile from Otterton which would perhaps indicate a site not far from the modern Clamour Bridge. The 'main road' since pre-medieval times ran from Peak Hill to Woodbury Castle encampments by way of Otterton from whence travellers could continue their journey by boat or ship. This may have been a great relief as the deep mud of Devon roads was notoriously difficult to traverse. The early prosperity of the settlements from which Otterton and East Budleigh have developed came from their proximity to the water's edge. They enjoyed the opportunities and suffered from the vulnerability that is inherent in such a position. Budleigh and Ottermouth were busy places importing iron, canvas, wax, glass, pottery and French wine and exporting wool, cloth and tin. Many of the local people depended on the river and sea for their livelihood.

Some idea of the type of ships which may have sailed from the Otter estuary can be gained from a study of a bench-end in East Budleigh church (6) which was carved in the early sixteenth century. This carving, possibly commissioned by a wealthy merchant or shipowner, shows a ship on water, a tiny figure in the rigging and a gateway in the background.

At Otterton there are several large old farmhouses bordering the street, a small village green with chestnut trees and a water-powered flour mill (7). The mill was recently painstakingly restored and brought to life again as a working mill grinding corn. Visitors can explore the building, see the mill working and purchase flour milled there. Lacemaking, a traditional craft of the area, is featured in an exhibition and sometimes a lacemaker is at work in one of the upstairs rooms. The ground floor is occupied by a craft shop. There are small workshops in the courtyard where baking, pottery and woodworking take place. Beside the car park is a small restaurant with indoor and outdoor seating.

In the fourteenth century the prosperity and the importance of the settlements in the Otter estuary began to decline when a pebble spit started to form across the estuary's mouth. Strenuous efforts were made to keep a channel clear of the offending shingle

and sandbanks so that sea-going ships could still operate, but to no avail; the estuary became silted up. By 1565, when lists of all possible places where goods might be imported or exported were drawn up for customs purposes, Budleigh and Ottermouth were no longer mentioned. Diminished river traffic continued for some time after the pebble spit had formed, though smaller boats replaced the sea-going ships.

In 1719 stone from Beer quarries was landed at Budleigh for the building of the nonconformist chapel at Lower Budleigh. Both the chapel and Vicar's Mead, the attractive thatched former vicarage of East Budleigh, feature in smugglers' tales of this area. Navigation was difficult as there were frequent floods and the course of the river sometimes changed after storms. In 1812, to confine the river, the local landlord, Lord Rolle, had an embankment built. Rock from the east side of the river was used to build the embankment on the west side. It now supports hawthorn, blackthorn, ash, oak and an 'orchard' apple tree which form an effective windbreak for walkers and a natural hide for bird-watchers. Alluvial silts collected to the landward side of the pebble spit and the acreage of salt-marsh and water-meadow gradually increased. Drainage ditches were dug and hedges grew on some of the banks between the fields. In more recent times some of the land beside the road to White Bridge and South Farm was used for a while as a rubbish tip for Budleigh Salterton but it has now been levelled and planted with trees. On the seaward side of the reclaimed tip a small freshwater marsh has developed fed by the stream flowing from Kersbrook. A seat has been provided so that walkers may pause awhile to watch and listen to the many insects and birds which populate this small but important piece of wetland habitat.

In recent years there have been fierce storms when it seemed that the elements might achieve what man had earlier failed to do and recreate a navigable estuary by sweeping away the pebble spit. But modern efforts to anchor the shingle have proved adequate to withstand the movement of the pebbles.

The shingle spit, sand, mud and silt formed new habitats for wildlife. Initially the instability of these new areas would have prevented the establishment of plant communities, but algae would quickly have formed in which other plants could grow and invertebrate life within the deposits to the landward side of the spit would have been able to establish itself. With increased stability, primary plant colonizers of salt-marsh would have a chance to develop in the mud above the high-water level of neap-tides. Other species could then follow as water movement slowed down and silt was deposited. The clarity with which such salt-marsh development is observable in the Otter estuary has made it a popular venue for educational groups. Its importance as an example of salt-marsh has been acknowledged and it is now leased to the Devon Trust for Nature Conservation by Clinton Devon Estates and is managed as a nature reserve.

On the shingle spit, sea beet forms clumps with dock and an occasional plant of rock sea lavender. Rock pipits calling 'tseep-eep',

Estuary salt-marshes, seen from Otterhead

together with the 'tchizzik' of pied wagtails, will almost certainly be heard at all times of the year as these birds eagerly hunt insects among the plants, along the tide-line and on the mud of the estuary. In spring and in autumn they may be joined by parties of grey and yellow wagtails which migrate to and fro by way of the Otter valley. All these birds make good use of the water's edge, feeding on the many insects which congregate there.

It is worthwhile to pause on the shingle spit to look and listen. If the wind is strong a postion below the ridge can provide a sheltered spot. On the seaward side, cormorants may be seen on the rocks of Otterton Ledge with outstretched wings drying off, or diving from the surface of the sea as they fish. The black-and-white plumage and the bright orange-red bill and pink legs of the oystercatcher stand out as they search the seaweed for mussels. Four common species of gull may be sitting preening or seemingly asleep on the shore or the rocks. Terns – sandwich, common and little – migrate along the coast here, often fishing as they go. In very stormy weather some may be seen huddled on the mud to the landward side of the shingle spit waiting to continue their journey. Waders, duck and ocean-going sea birds also use the rocks of Otterton Ledge as a resting and feeding place when on passage. These birds too may seek shelter in the river channel. It is not only salt and fresh water that meet there but also plants, birds and other animals which are adapted to, and take advantage of, this special habitat. Man is not alone in exploiting the estuary.

A section of the Coast Path runs from the north-east corner of the car park to White Bridge along the edge of the nature reserve. At the car-park end there is an information board which illustrates some of the wildlife of the estuary. Primary colonizers of the salt-marsh can clearly be seen in the mud above the high-water level of neap-tides; these are annual glasswort, annual sea-blite and common salt-marsh grass. In addition, there is considerable growth of cord grass, which

can grow below the high-water level of neap-tides. It is a vigorous grower and can become dominant unless checked. Above the tide-line the blue-green leaves of sea couch grass catch the eye. In places it is mixed with the common couch. In early spring the white flowers of scurvy grass spread over the salt-marsh, to be followed later by clumps of sea-pinks and, with the cessation of grazing on the marsh, sea lavender. Less brightly coloured and less noticeable to any but the close observer, though equally important to the salt-marsh, are lesser salt-marsh spurrey, greater sea spurrey and sea milkwort, all of which have small pink flowers. Sea plantain, sea purslane and sea arrow grass make up the bulk of the foliage, the grey-green foliage of the purslane giving the marsh a characteristic hazy appearance, particularly at the edges of the creeks.

Zoologically the mud-flats are rich in invertebrate life. Myriad holes in the wet mud mark the entrances to the burrows of *Corophium*. The tiny fan-shaped patterns in the mud show where antennae have swept detritus into waiting mouths. A patient watcher may see the green worm, *Nereis diversicolor*, shrimp-like creatures and *Sphaeroma*, looking like woodlice, scurrying about in salt-pan pools or being swept along by the current in the river. Many small shore crab are stranded on the tide-line and the tiny snail, *Hydrobia*, which is eaten by shelduck, can be found among the vegetation.

The footpath is bounded by a hedge on the east and a steep bank down to a drain and fields to the west. About a quarter of a mile from the sea, plants of the common reed, *Phragmites*, begin to appear in and around the drain, while on the other side of the path, screened by a hawthorn hedge and some ash trees, the salt-marsh gives way to an expanse of bare mud or water which was once the main channel of the river but is now a side creek flooded only at high tide. A little further along, the drain broadens out into a small freshwater marsh with a reed bed and several marsh plants – bold stands of purple loosestrife, yellow flag iris, ragged robin, teazel, marsh marigold, and meadowsweet. Beside the path various vetches flower in their season and there are also handsome plants of comfrey, soapwort, hemlock and hemlock water dropwort.

In September the bright yellow flowers of common fleabane brighten the rough damp ground near White Bridge. Dragonflies and damselflies can frequently be seen in addition to several species of butterflies and many other invertebrates. The plants of the saltwater and freshwater marshes were more widespread in the past in the estuary and were used for domestic, medicinal and commercial purposes. Rushes were strewn on floors and made into baskets and hats. The medicinal properties of scurvy grass, meadowsweet and comfrey have long been recognized, but the attributes of many still await discovery. Purple loosestrife was sometimes used in place of bark in the local tannery. Soapwort, the fuller's herb, was used in the production of woollen cloth in fulling mills and may have been so used in the mills of the Otter Valley. The use of teasels in the wool trade is recorded on one of the carved bench-ends in East Budleigh church.

The birds, which take advantage of the abundance of food in the varying habitats, change with the seasons. The valley and the estuary have featured for many years in ornithological records, not so much for the numbers of birds, as for the variety of species which have used the areas. It has long been a migratory route used in autumn and spring by waders and wildfowl, hirundines, warblers and other passerines. Birds of the water's edge feed, rest and preen in the estuary, on the mud-flats, the salt-marsh, the hedges, the drainage dykes and the fields. The nineteenth-century embankment and twentieth-century drainage have reduced the area of both salt-marsh and freshwater marsh in the lower valley and estuary, so few birds breed here now. However, in most years mute swan, mallard and shelduck young can be seen in the river, as well as moorhen chicks and the young of sedge warblers, reed warblers and reed bunting in and around the reed beds.

In autumn, winter and spring, redshank, greenshank, common sandpiper, dunlin, ringed plover, grey plover, lapwing, curlew and snipe are frequently seen. Birds can often be observed at close proximity as they feed on the mud near the footpath. As numbers of any one species tend to be small, and varied food plentiful, it is often possible to see several species close together and to make detailed comparisons to aid identification. Mallard are present throughout the year, joined from autumn to spring by shelduck, teal, wigeon and a large gaggle of Canada geese. Little grebe, cormorant and red-breasted merganser can be seen fishing the river, out of the breeding season. They all dive from the surface of the water and remain submerged for a considerable time, popping up again in a different place. Grey heron, probably from Powderham heronry on the west bank of the Exe estuary, seem to find plenty to eat in the river and marsh dykes. They stand motionless in the shallows until with a swift jab they catch the prey in their long beaks. The commonest gulls, herring and black-headed, are sometimes absent from the estuary but may be seen as a white cloud following the plough in nearby fields. Lesser and greater black-backed gulls use the estuary in small numbers, and occasionally common gull and kittiwake are seen. Kingfishers fish in the river, the dykes and the creeks, making a brilliant splash of colour in the winter scene. In the salt-marsh, flocks of starling, greenfinch and linnet are hard to see as they feed on the ground but in the air or the hedges form excited twittering flocks. Reed bunting, goldfinch, chaffinch and tits which frequent the pastures, low vegetation and hedges are joined during autumn and spring migration by warblers, wheatears, chats and wagtails. The noisy songs of the sedge and reed warbler can be heard all summer from the reed beds over which large numbers of swallows, swifts and house martins hunt for insects. In winter, water rail may be sighted. With so many birds it is not surprising that birds of prey also visit the reserve area. Starlings gathering in a hedge before roosting suddenly rise in a mass with strident voices as a sparrowhawk streaks among them. A kestrel hangs motionless in the air before dropping like a stone. Occasionally owls quarter the marshes and fields.

Heron

Raptors are not the only hunters about. Mink seem to be established in the area and are frequently seen in the river and on the banks. Otters had disappeared as a breeding species before the mink arrived, though in the between-wars period there was still otter-hunting in the valley. Foxes forage on the marsh and their tracks can be seen in the mud at the river's edge. Grey squirrel inhabit the pine trees and have been seen to carry large cabbage leaves aloft when disturbed by walkers on the Coast Path.

Just as man's life at the water's edge was dramatically changed by the build-up of the shingle spit from the fourteenth century onwards, so the life of plants and animals in the estuary is continually altered as a result of man's activities. Changes can happen so rapidly with the use of modern technology that there may be no time to adapt. Fortunately the land in the area is in the single ownership of Clinton Devon Estates so it can be managed as a whole. For generations the valley has been a peaceful place, adapting to change but keeping its character and the inhabitants their concern for the wildlife. The Otter estuary provides a unique public amenity for those who find relaxation in the appreciation of the natural environment and in unravelling the evidence of man's use and care for the estuary's resources.

This chapter was prepared by Miss B. P. R. Primmer on behalf of the Otter Valley Association. The Association was formed in 1979 and now has about 800 members. For the public benefit the Association seeks to secure the preservation, protection, development and improvement of features of the built and natural environment of the five civil parishes in the Lower Otter Valley, from Newton Poppleford seaward to Budleigh Salterton. The Association may be contacted c/o 6 Marine Parade, Budleigh Salterton, Devon EX9 6NS.

3

Exe Estuary

The Exe estuary is not only the largest estuary in Devon but also the most outstanding for wildlife. In earlier times it also had considerable local importance for maritime trade but this trade has largely disappeared today.

The shallow entrance to the estuary and its winding channel have always posed difficulties for navigation but in the Middle Ages small vessels were able to reach the upper estuary and Exeter, although even then Topsham was preferred as an easier place to get to.

In the sixteenth century there was a regular anchorage off Dawlish Warren where vessels waited for wind and tide. Trade increased, especially the wool trade, with Topsham holding a prime position and attracting Dutch shipping agents. Relics of this period can be seen in the Dutch-style houses along the river.

A major factor in Topsham's importance was the blocking of the estuary at Countess Wear. This dates from 1284, when Countess Isabella of Devon began the work which was later completed by Hugh Courtenay, Earl of Devon. For more than two centuries Topsham was the chief port until the merchants of Exeter built a canal to by-pass the Wear. The Exeter Ship Canal was the first of its kind and was later extended to Turf Lock, more than two miles downstream.

Today the canal is an ideal viewpoint for the upper estuary, with its towpath providing good access along the western side. The maritime trade declined with the arrival of deeper-draughted vessels in the late nineteenth century and the development of Exmouth Dock (1) which took larger vessels out of the estuary.

The other factor to have a great effect on the Exe was the arrival of the railway. The flat land alongside the estuary attracted the railway companies and a line was built on either shore. Artificial embankments along the river were created for the rail tracks, but these have helped to restrict other developments which might have affected the estuary's now widely acknowledged wildlife interest.

The Exe is one of the most outstanding wildlife areas in south-west England. The reasons for this are complex but perhaps the most obvious is its size. Unlike many of the other estuaries further west, it is very open and shallow so that its mud-flats and sandbanks are wide and extensive. At its seaward end it is almost one and a half miles wide and from Countess Wear to Dawlish Warren it is more than six miles long. Throughout this area the mingling of fresh and salt water creates a rich environment for millions of organisms. Some are too small to be seen with the naked eye (nematode worms can exceed ten million per square yard), others, such as ragworms, can be up to fifteen inches long. Small snails, molluscs, shrimps, etc., all form part of the food-chain which, in turn, supports tens of thousands of fish and birds.

Exeter Canal

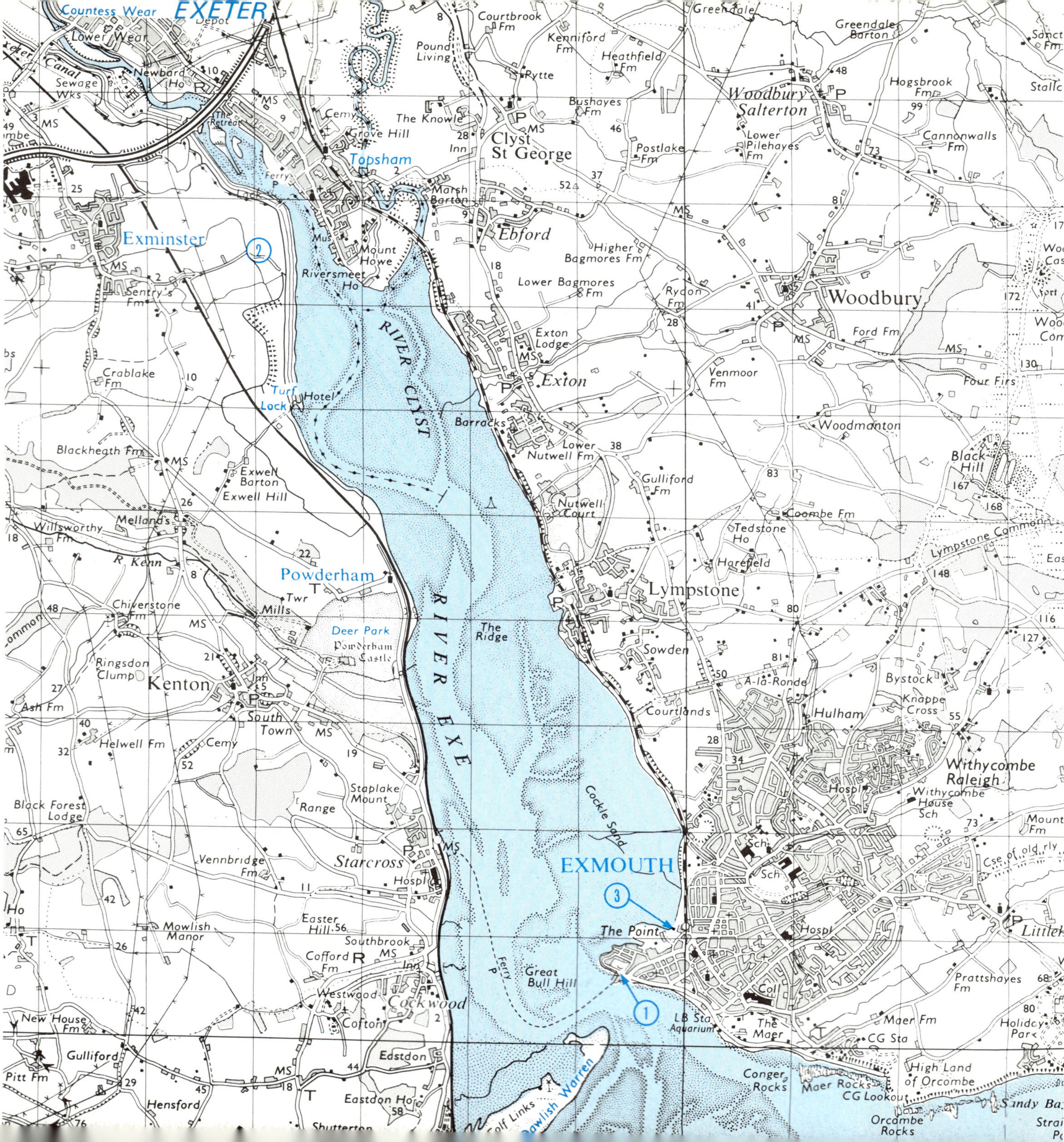

EXETER
Countess Wear
Lower Wear
Depot
Canal
Sewage Wks
Newhard Ho
The Retreat
MS
Cemy
Grove Hill
The Knowle
Topsham
Ferry
Exminster
Sentry's Fm
Mus
Mount Howe
Riversmeet Ho
RIVER CLYST
Crablake Fm
Turf Lock
Hotel
Blackheath Fm
Exwell Barton
Exwell Hill
Willsworthy Fm
Melland's
R Kenn
Powderham
Twr Mills
Chiverstone Fm
Deer Park
Powderham Castle
Ringsdon Clump
Kenton
South Town
Ash Fm
Helwell Fm
Cemy
RIVER EXE
The Ridge
Black Forest Lodge
Range
Staplake Mount
Starcross
Hospl
Vennbridge Fm
Mowlish Manor
Easter Hill
Southbrook
Cofford Fm
New House Fm
Westwood
Cofton
Cockwood
Gulliford
Pitt Fm
Hensford
Eastdon Ho
Shutterton
Golf Links
Dawlish Warren
Pound Living
Courtbrook Fm
Kenniford Fm
Heathfield Fm
Pytte
Bushayes Fm
Clyst St George
Inn
Marsh Barton
Ebford
Postlake Fm
Higher Bagmores Fm
Lower Bagmores Fm
Exton Lodge
Exton
Barracks
Lower Nutwell Fm
Nutwell Court
Gulliford Fm
Rydon Fm
Venmoor Fm
Woodmanton
Coombe Fm
Tedstone Ho
Harefield
Lympstone
Sowden
A-la-Ronde
Courtlands
Hulham
Bystock
Knappe Cross
Withycombe Raleigh
Hospl
Withycombe House Sch
EXMOUTH
The Point
Great Bull Hill
Ferry
Aquarium
LB Sta
The Maer
CG Sta
Conger Rocks
Maer Rocks
CG Lookout
High Land of Orcombe
Orcombe Rocks
Woodbury Salterton
Hogsbrook Fm
Cannonwalls Fm
Woodbury
Ford Fm
Four Firs
Black Hill
Lympstone Common
Prattshayes Fm
Maer Fm
Holiday Park

The ebb and flow of the tide moves the sediments of the estuary bed so that sands and muds separate. The muds support a somewhat richer biomass and in the lower Exe provide the growing conditions for one particular plant, *Zostera marina* (eelgrass). The extensive areas of this plant which grow below the tide-line are an important reason why the Exe supports the richest wildfowl populations in Devon.

At the edge of the tide-line the growth of several plants begins to consolidate the sediments and form salt-marsh. The main colonizer is *Spartina* (cordgrass), a plant which has produced hybrids and crosses from introductions. There are, however, indications that the vigour of some stands of these plants is beginning to wane. The Exe is fairly exposed to winter storms so that areas of sheltered salt-marsh are limited, but where they occur a variety of interesting plants are found.

The extensive dune system of Dawlish Warren, stretching across the mouth of the estuary, is a mobile system, despite man's attempts to stabilize it. It provides another niche for a variety of plants and animals and, in particular, one of the most important high-tide roosts for the birds. Upstream the open Deer Park at Powderham and the grazing meadows of Exminster Marshes (2) are both important areas within the estuary's ecosystem, providing important additional habitats, feeding and breeding areas for the birdlife.

The Exe estuary and Dawlish Warren

Brent geese feeding on Exminster Marshes

Oystercatchers arriving at Dawlish Warren

Avocet

Wildfowl

The wildfowl of the Exe are most numerous in winter when dark-bellied Brent geese and wigeon visit the estuary in thousands to feed on the eelgrass and later the flooded pastures. Numbers of Brent geese now exceed four thousand and wigeon numbers are often even higher. Other winter visitors include thousands of mallard and teal plus significant flocks of pintail and mute swans.

Further out in the estuary are diving ducks: eiders and scoters in the mouth of the river, red-breasted mergansers both in the mouth and further upstream.

Waders

Numbers of waders on the Exe often exceed twenty thousand (a number considered to indicate international importance). The sheer variety is impressive, with over twenty species usually represented. At low tide they are spread widely over the open mud-flats and shellfish banks, at high tide they congregate in certain undisturbed areas. Although the areas may overlap, there is a marked tendency for certain species to frequent certain parts of the estuary, obviously relating to their feeding habits. Downstream at Exmouth and Dawlish Warren the main species are sanderling, turnstone, oyster-catcher, bar-tailed godwit, knot, grey plover, ringed plover and dunlin. Further upstream most of these species occur but often more numerous are redshank, curlew and black-tailed godwit with important flocks of greenshank and avocets. The black-tailed godwit is possibly the most important wading bird in national terms. Average peak numbers of this long-billed wader can reach six hundred, representing 1.7 per cent of the north-west European population, a very significant proportion.

Many of these birds use the Exe as a migratory stopover on autumn and spring passage. Ringed plover and curlew occur in noticeable peak numbers in autumn. Others, like oystercatchers, come in late summer and spend the whole winter on the Exe, returning year after year. The destination of individual birds on spring migration can vary by thousands of miles: sanderling heading for Greenland; whimbrel and black-tailed godwits for Iceland; knot for the Arctic Circle. The Exe then is like a migratory crossroads, an essential refuelling stop which has provided a haven for thousands of years.

Among the more recent changes in this ancient pattern has been the increase in numbers of avocets. These elegant black-and-white wading birds have gradually increased over the last decade until now more than a hundred visit the upper estuary each winter from November until February.

Migrants and breeding birds

Although peak numbers of birds on the Exe occur in mid-winter, the only quiet period is in late June and early July. At other times spring and autumn passage bring thousands of birds to the estuary. April sees the first spring migrants: martins and swallows over the meadows and reed-beds, sandwich terns on the sandbanks of the outer estuary and the first wheatears at Dawlish Warren. Gradually, more and more migrants appear, including other waders en route further north. The breeding birds begin their activity early, herons at Powderham Park displaying even in February. The migrant breeding birds include the colourful yellow wagtails which arrive from Africa to breed on Exminster Marshes. Here also are a few breeding lapwing and redshank, the only waders which remain to breed. Breeding wildfowl normally consist of mallard and shelduck but occasionally include a pair of the somewhat scarce garganey, a summer visitor from Africa.

The upper estuary is fringed by reeds which provide habitat for hundreds of reed and sedge warblers whose harsh song fills the air from May onwards. More musical is the powerful song of the Cetti's warbler, a newly arrived species which is a resident more often heard than seen.

Autumn migration begins early so that by July returning birds can be seen still in summer plumage. Terns at Dawlish Warren can include up to five species, including one or two elegant roseate terns. This is also the time of year when a visiting osprey may pay a call, diving from above the incoming tide to catch a grey mullet. Powderham is often chosen by these attractive birds of prey as a roost site. Autumn also brings the first wintering peregrine falcon to chase the flocks of waders. Some migrants arrive only in harsh winters; Bewick's swans and white-fronted geese can then be found on the marshes. Short-eared owls are more regular but, even so, their numbers can vary from year to year.

Plants and insects

Around the estuary is a wealth of other wildlife. The dune system of Dawlish Warren supports a variety of plants from the introduced tree lupins and evening primroses to the native spring and autumn orchids. Especially famous for the diminutive Warren crocus which flowers in April, the area has many other species of interest. Upstream the ditches and streams of Exminster Marshes support an attractive aquatic flora and in mid-summer are the haunt of dragon-flies and damselflies. The estuary also attracts and supports a wide range of other interesting insects, including butterflies such as painted ladies and clouded yellows, migrants from the continent. The day-flying Jersey tiger moth is a local speciality.

Conservation

The importance of estuaries such as the Exe is now recognized by a host of special designations. The estuary is graded as internationally important for certain species of waders and wildfowl and for total numbers. It has government designation as a Site of Special Scientific Interest and is also being proposed as a Special Protection Area under an international agreement (Ramsar Convention). In addition to these designations there are several nature reserves and possibly more in the pipeline. These official titles do not remove the need for extreme care in developments associated with the estuary and, in particular, the need for zoning of recreational activities. Watersports in the estuary are rapidly growing in popularity and the need for an overall approach to such activities is becoming more apparent year by year. This is why the more people there are who appreciate and care for the wildlife of this exciting and unique area, the better will its future be secured.

Low tide on the Exe estuary

Viewing the wildlife

To see the whole range of the Exe estuary wildlife, which is continually changing with the tides and seasons, many visits are necessary. It is specially important to time your visit in relation to the tide. At low tide the waders and wildfowl are spread out right across the mud-flats so that many are out of the range of binoculars. As the tide begins to rise so the birds congregate on the areas still uncovered, until at last they are forced by the rise in water-level to move to roosting areas until the tide drops once again. These roosting areas are chosen because of their safe, undisturbed nature and on the Exe estuary there are several places where the flocks of waders and wildfowl can be viewed at a distance without disturbing them. These are:

Dawlish Warren Local Nature Reserve High-tide roost at far end of golf course where two-tiered hide is available. Access along beach and past end of golf course.

Powderham Park View from road between railway and parkland.

Exminster Marshes and fields View from sea-wall running from Exminster (Swan's Nest Road) to Turf Lock. Access also from Topsham pedestrian ferry.

Exmouth: Imperial Road car park **(3)** Good from November to March, especially for wildfowl.

Powderham Park and Exminster Marshes are particularly good when flooded in winter.

During most winters special cruises are arranged by the Royal Society for the Protection of Birds to see the avocets and other birds of the estuary. (For details, contact the Regional Office, 10 Richmond Road, Exeter.)

At Dawlish Warren, guided walks take place throughout the year. More information on these is available from Teignbridge District Council, Planning Department, 32 Courtenay Street, Newton Abbot, Devon.

The estuary from Exmouth docks

This chapter was prepared by Stan Davies, the Regional Officer (South West) of the Royal Society for the Protection of Birds, on behalf of the Exe Estuary Conservation and Study Group. The Group includes representatives of a range of scientific and conservation bodies and seeks to secure the long-term conservation of the estuary by encouraging interest in its scientific and wildlife value.

The Royal Society for the Protection of Birds aims to encourage the conservation of wild birds and their habitats. It has 375 000 members and its South West Regional Office is at 10 Richmond Road, Exeter, Devon EX4 4JA.

4

Teign Estuary

The River Teign has one of the smaller of Devon's estuaries but it may well be regarded as among the loveliest. From where it joins the sea at Teignmouth, the estuary extends inland almost to the outskirts of Newton Abbot. It is some four miles in length and at its broadest point is nearly half a mile wide.

The estuary is surrounded on all sides by beautiful scenery designated as Great Landscape Value and Coastal Preservation Areas. Dominating the eastern, seaward end at Teignmouth stands the bold and striking red sandstone Ness headland (1), two hundred feet high, while to the west, far behind Newton Abbot, there is a matchless background of the highlands of the Dartmoor National Park.

The River Teign rises in a high bog area of northern Dartmoor to the south of Okehampton. Throughout its length it is a clean and attractive moorland and rural river. It has two main tributaries both of which also rise on Dartmoor: the Bovey, which runs into the Teign some three miles north of Newton Abbot, and the Lemon, which, after flowing through Newton Abbot, partly in a culvert, joins the Teign shortly before it enters the estuary.

A small pretty stream, the Archbrook, runs directly into the estuary on its southern shore at an attractive creek beside the Shaldon–Newton Abbot road (2).

The Teign Estuary derives much of its charm from its un-developed character. Along its shores there is little more than scattered building until it reaches the town of Teignmouth (population: 12 000) and, on the opposite bank, the village of Shaldon (population: 2000) which includes the adjacent small settlement of Ringmore. These are pleasant residential and holiday resorts linked by the quarter-mile long Shaldon Bridge (3). In addition, a small motor ferry (for foot passengers only) runs a frequent service between Teignmouth and Shaldon river beaches near the harbour entrance.

The attractive small natural harbour at Teignmouth contains a busy commercial docks area, with ships of many nationalities and up to some two thousand tons in size entering and leaving almost every

The Teign estuary from Bishopsteignton

day. It has been an active port throughout recorded history and in particular during the seventeenth, eighteenth and nineteenth centuries when there was an important trade with Newfoundland. Each summer many Teignmouth vessels crossed the Atlantic to the prolific fishing grounds off Newfoundland. Manufactured goods were taken out to the settlers and salted codfish brought home. Nowadays, a main trade is the export of ball clay mined in the Bovey Basin north of Newton Abbot, while animal feedstuffs are a chief import. A variety of other cargoes are handled from time to time, including timber, building materials, fertilizers and coal.

Ideford
Olchard
Luton
PH
Moor Brook
Well Covert
ROMAN ROAD
Hestow Barton
Lindridge Park
Humber
Whiteway Barton
Lindridge Hill
Ashwell
Cemy
Bishop's Palace (remains of)
Wolfsgrove
Rydon
Murley Grange
Wood
Kingsteignton
Ash Hill
Bishopsteignton
Chapel (rems of)
Wear Fm
MS
Greenhill
Ware Barton
RIVER TEIGN
Coombe Cellars
Nethert Ho
Combeinteignhead
Netherton
Buckland Barton
Works
Arc Brook Br
Ringmore
NEWTON ABBOT
Buckland
Milbe
Newtake
Haccombe
Haccombe Ho
HACCOMBE WITH COMBE
Charlecombe
No Man's Land
Lower Rocombe Fm
STOKEINTEIGNHEAD
Stokeinteignhead
The Beacon
Forches
Aller Park
Fort
Sand & Gravel Pit
Aller
Middle Rocombe
Higher Rocombe Barton
Higher Gabwell
Lower Gabwell
Devon South Coast Path
Coffinswell
Court Barton
COFFINSWELL
Daccombe
Great Hill
Sladnor Park Ho
Maidencombe
Kingskerswell
Schs
Bell Rock
Rock Ho
Fluder
Hotel
Barton
Watcombe
Watcombe Head
Little Haldon
Golf Course
Chapel (rems of)
Castle Dyke Tumulus
Smallacombe Fm
Lidwell
Southwood Fm
Venn Fm
Coombe
CH
Luscombe Castle
Aller Fm
Cemy
Oaklands Park
Westbrook Fm
Holcombe Down
Holcombe
MS
Sprey Point
Coryton's C
Horse Cove
The Parson and
DA
Breakwa
CG L
Mus
Sch
TEIGNMOUTH
Hospl
The Salty
Pier
Shaldon
The Ness Ho
Ferry
Bundle Head
Labrador Bay
BABBACOMBE BAY
Mackerel Cove
Blackaller's Cove
Watcombe Head

There is an active local fishing industry with a small fleet of inshore trawlers, as well as salmon seine boats which fish in the estuary. There is also an important mussel and oyster fishery.

Recreational pursuits include yachting and boating. Competitive sailing matches are organized by the Teign Corinthian Yacht Club and Shaldon Sailing Club. Power-boating and water-skiing take place, above the bridge only, the organizing body being the South Devon Water Sports Club.

Angling is another popular activity, the main species caught in the upper estuary being flounder and mullet, while around the harbour mouth there are bass, mackerel, pollack, conger-eel and wrasse. No permits are required for these fish. For those wishing to join an organization, the Teignmouth Sea Angling Society provides facilities and holds periodic competitive events.

Many parts of the estuary are suitable for swimming if care is taken to keep clear of moving motor boats and tidal currents. Official warning notices should be heeded.

A feature of the lower estuary and harbour is the large central shingle-bank known as The Salty. Formed over the ages from material swept in from the sea, at low water this dries out into an extensive area of firm beach. The wide tide-washed expanse is a pleasant local amenity where children and holiday-makers can wander and explore in safety and anglers can collect bait.

Teignmouth and Shaldon are the only permanent settlements actually on the estuary shore, but in the immediate hinterland to the north there is the attractive and historic village of Bishopsteignton, while near the southern side there are small and ancient Coombeinteignhead and Netherton, scarcely more than hamlets. There are holiday camps beside the estuary, both in the Shaldon area and also on the northern bank near Bishopsteignton.

The estuary's northern shore has its access restricted by the British Rail Western Region main line which runs beside it, this stretch being regarded as among the line's most picturesque sections. Above Shaldon and Ringmore the southern shore is largely undeveloped and in natural condition. Much of it is in private ownership so that here too access tends to be restricted although it is possible to walk along the beach when the tide is out.

To see the whole of the estuary properly it is best to go by water. Boats can be launched from a number of places around the estuary and the harbour area. Local watermen hire out boats from Teignmouth and Shaldon river beaches and they also run organized trips up the estuary.

Sheltered and safe as the estuary is, a newcomer is advised to go at first with someone familiar with it or at any rate to have a word beforehand. Much of the estuary is shallow and no one likes a long and possibly chilly spell stuck on a mudbank waiting for the tide to return. The bridge of course needs care; slack water is the best time for a novice to pass under, but if it is necessary to do so when the current is running at all strongly, a boat should make for the Teignmouth side where the main channel is and go through the middle of the slightly highest and widest arch in midstream, keeping well clear of the pillars.

The present steel and concrete bridge dates from 1931 when it replaced a wooden structure erected in 1838. The stonework approaches on either side are those of the very first Shaldon Bridge which was built in 1827 but which collapsed prematurely a mere eleven years later as a result of the timber being bored into by ship-worm. Baulks of timber from the wooden bridge, dismantled in 1931, may still be seen doing duty as 'mooring posts' on the beach at Teign View Place, Teignmouth and Marine Parade, Shaldon and may be identified by the numerous copper nails driven in to deter ship-worm.

Aerial view of the Teign estuary

Shaldon Bridge

The old bridge was reputed to be the longest wooden bridge in England at that time and the second longest in Europe. Formerly owned by a private company which charged a toll, the bridge was freed when it was taken over by Devon County Council in 1948.

In the area around the harbour, a landmark of interest is Bitton House (4), an imposing Georgian mansion not far from the Teignmouth end of the bridge. It is now owned by the District Council and used as offices, while the pleasant grounds form a popular public park. The house once belonged to Admiral Pellew (later Lord Exmouth), the Naval hero famed for his successful siege of Algiers in 1816, and two cannon captured on that occasion still stand sentinel outside the building. Beside the house there is a graceful antique orangery erected by Admiral Pellew. This had latterly fallen into disrepair and had been threatened with demolition, but is now being restored by the joint efforts of the District Council and local conservationists.

Opposite, on the Shaldon side, is St Peter's Church (5) built in 1900. The interior is very beautiful although the exterior lacks the balanced appearance given by a spire. It had been intended that one should be erected but this was found impossible on the sand foundations.

In former times much of the commerce of the port was handled at Shaldon which took an active part in the Newfoundland trade. The cargo vessels, which of course in those days were small sailing ships,

Bitton House, Teignmouth

were berthed alongside wooden jetties built out from the shore. The stonework approach to one of these is still in existence at the Clipper Cafe which occupies the site of a former large coal store. Summer holiday-makers enjoy ice creams at sunshaded tables where once coal was laboriously unloaded by hand.

Further seawards, facing Shaldon's Marine Parade, the curiously designed round-ended building, now three dwellings, was originally a warehouse associated with the Newfoundland trade. Formerly named Gempton's Cellar, it was converted into dwellings in 1821. It is believed to be one of the largest existing cob structures in this part of the country.

Above the bridge, another landmark on the Shaldon side is quaint castle-like Ringmore Towers (6). Despite its appearance of antiquity this is in fact a comparatively modern 'folly' built at the turn of the century on the site of an old cottage.

Nearby on the Embankment, at the King George's Field recreation ground, there was once a shipyard of some importance which handled quite large vessels, until reputedly it was placed at a disadvantage by the construction of the bridge. Nowadays its sole commemoration is in the name of the adjoining Shipwrights Arms Inn and, beside the Ringmore Road, a solitary red sandstone gate pillar, formerly one of three at the yard's main entrance.

Further on, facing down-river, lies Ringmore Strand with its attractive period dwellings little altered from their appearance in an eighteenth-century watercolour sketch in the possession of a local resident. This is a Conservation Area, as is most of the main village of Shaldon and its riverside waterfront.

Behind the Strand is tiny St Nicholas' Church (7), dating from Saxon times. On an outside stone at its north-east corner can still be seen the faint outline of a very small cross, believed to have been engraved by a Crusader. It is said that before leaving for the Holy Land a Crusader would cut one half of a small cross in the wall of his home church; if he returned he completed his cross.

Before St Peter's was built, the only church in Shaldon was St Nicholas'. At one time it had been necessary to add a large out-of-character extension to accommodate the congregation. When no longer required this was removed and the church restored to its original form. The old church retains its popularity and is well filled every Sunday. A few years ago a proposal for its closure had to be dropped because there was such strong opposition by residents.

Beyond Ringmore and the nearby holiday camp lies the undeveloped estuary with low wooded cliffs and little rocky beaches, perhaps much the same as it looked in the medieval days of that long-ago Crusader. Attractive at any season, the best time of all to come is in the spring when the wild cherry trees are in bloom, a sight of unforgettable beauty overhanging the water.

Past Archbrook Creek and further woody cliff and beach, stands the ancient Coombe Cellars Inn (8). As well as being a popular social venue, this is also the power-boating and water-skiing centre.

On the northern bank, almost opposite Coombe Cellars, the promontory of Flow Point (9) juts out clear of the railway line and is a

Archbrook Creek, with Coombe Cellars Inn in the distance

centre for boating and angling. Most of the rest of the northern shore being taken up by the railway, the other interests here are virtually limited to the mussel and oyster fishery and the collection of bait.

It is possible to go up well beyond Coombe Cellars and Flow Point, past the Netherton area and on almost into Newton Abbot. However, the higher reaches of the estuary are less suited to boating, although they have been much improved in recent years since the new Newton Abbot sewage-treatment plant at Buckland has achieved considerably cleaner water.

In these higher reaches the huge reed banks of Hackney Marsh stretch across the estuary, which was spanned a few years ago by the massive viaduct carrying the A380 Exeter–Torbay road **(10)**. A fine view of the estuary can be seen while driving across, but it is regrettable to note the extent to which Newton Abbot's industrial outskirts are encroaching into the reedy wetland and the habitat of the river birds. In this area the Devon Wildfowlers Association have an agreement with the Devon Trust for Nature Conservation that the south side of the estuary should be managed as a bird sanctuary and on the north site a limited amount of licensed shooting should be allowed during the winter season.

Aerial view of the Dart estuary

Salcombe from Mill Bay on the Kingsbridge estuary

The estuary is important for its birdlife, especially waders, the extensive mud-flats providing feeding grounds. Mallard and shelduck nest along the estuary and adjoining land and there is a breeding colony of herons at Netherton Woods that uses the estuary as feeding grounds. Marsh-loving birds such as sedge and reed warblers occur and mute swans, Canada geese, and a variety of sea birds are found. Various more unusual species visit the estuary on spring or autumn migration and during cold spells in winter.

At the head of the estuary there were formerly two canals running inland, the Stover Canal and the shorter Hackney Canal. A main use of both was the transport of ball clay by barge to Teignmouth where it was transferred to seagoing cargo ships. Granite quarried near Hay Tor on Dartmoor was also brought out along the Stover Canal, having first been drawn by horses along a tramway made of granite blocks to the head of the canal at Stover, some three miles from Newton Abbot. The granite was shipped from the specially erected New Quay at Teignmouth and the uses to which it was put included the construction of various important London buildings, among them the British Museum and the National Gallery. There was also an import trade up the estuary, a variety of goods, including coal, being taken by canal to supply the country areas. One of the last of the river barges, the *Two Brothers*, lies in the mud at Archbrook Creek, and although derelict, may be seen to have been a well-built little vessel.

The canals have long been disused, although parts are to be seen in derelict condition. Near their junction with the estuary at Hackney on the northern bank there is still the ancient Passage House Inn (11) which formerly catered for the bargemen. With changing times it is now a popular social venue.

Also associated with this period of the estuary's history is an imposing line of tall warehouses close to the river, off Quay Road at Newton Abbot. They are today used mainly as maltings, but the magnitude of these fine old buildings pays tribute to the former volume and importance of water transport.

Throughout its length the Teign Estuary is full of interest and fascination at every season for those of all ages and inclinations. It is so very much part of the way of life of this lovely locality that it is small wonder that visitors return over and over again and residents are loath ever to move away.

This chapter was prepared on behalf of Teignmouth and Shaldon Environment Society by the Hon. Secretary, Miss F. M. Atkinson. Formed in 1972, the Society has 400 members and its aims are to work for preservation, conservation and prevention of pollution in the area of Teignmouth, Shaldon, Bishopsteignton and the adjoining countryside, coast and estuary. The Society can be contacted at the Hon. Secretary's address: Cambria, 4 Marine Parade, Shaldon, Teignmouth TQ14 0DP.

The Society wishes to thank Teignbridge District Council for assistance and for information obtained from their Teign Estuary Study. Other sources of information for which the Society is grateful are History of Teignmouth by G. D. and E. G. C. Griffiths, Westcountry Harbour by H. J. Trump, and The Potter's Field by L. T. C. Rolt.

5

Dart Estuary

The tidal estuary of the Dart begins just above Totnes Bridge, from where it meanders for eleven miles to reach the open sea at Dartmouth. Curiously, the 'Water of Dart' from the Anchor Stone near Dittisham to the Western Blackstone (1) outside Dartmouth has belonged since 1333 to the Duke of Cornwall, now Prince Charles, who is entitled by a relic of feudal days to collect money from anyone who moors a boat in his water, or even builds a bay window over the edge of it. Most of the Duchy rights over the water have now passed to the Dart Harbour and Navigation Authority, who control all shipping using it and collect harbour dues.

The estuary is in fact a ria, or drowned valley, scoured out into a steep-sided valley by spring melt-waters during the Ice Age, then flooded as water-levels rose when the ice melted finally. Over the years, especially in the upper reaches, hundreds of tons of sediment have been washed down to form mudbanks, a rich source of food for birds, though making navigation hazardous at low tide for the unwary. Below Dittisham, where the wooded hills plunge down from over five hundred feet at one of the narrowest stretches of the river, the channel deepens and resembles a fiord.

The woodlands fringing the Dart today, which greatly enhance its beauty, are mostly the result of deliberate planting over the past hundred years by local landowners. Mid-nineteenth-century prints of the area show bare hillsides, and the tithe maps confirm that in 1841 few were tree-covered. Long Wood (2), south of Maypool, and now owned by the National Trust, is the exception in being an ancient oak wood, probably more than three hundred years old, which was once coppiced to provide tannin from the bark for tanning leather, and charcoal from the timber for smelting metals. Gallant's Bower, Dyers Hill (3) and Hoodown (4) were all before the late-nineteenth century just fields, whereas now they are covered by fine deciduous woods of oak, beech and sweet chestnut. All are now owned by the National Trust, under whose careful management their beauty will be preserved for the future by thinning out and replanting to provide more varied species as well as those already there. On the Kingswear side, near the harbour entrance, lies an unusual belt of Monterey pines – a species discovered only in 1830 in California – which appears to have been planted early this century. These trees survive well on this sheltered south-facing site, and in turn provide shelter for gardens behind, blending well with the local species.

From Noss downstream the river changes its character to become busier as Dartmouth and Kingswear appear, with two car ferries providing the first chance since Totnes for motor traffic to cross the Dart. Here in summer time will be hundreds of small sailing craft. Tripper boats take visitors round the harbour entrance, or to Torbay or Totnes. Two marinas cater for yacht owners and two yacht clubs organize regular races. The Royal Regatta is the occasion of the year at the end of August.

The mouth of the Dart, with Kingswear on the left

TOTNES
Longcombe
Lower Longcombe
Parliament
Motel
Inn
Sch
Goodrington Sands
Zoo
Clayland Cross
Clennon Hill
Windmill Hill Clump
Resr
Aish
Whitehill
Yalberton Tor
Yalberton
Works
Goodrington
Saltern Cove
Sharpham Barton
ASHPRINGTON
Sharpham Ho
Port Bridge
Homestead
Lower Well Fm
Sch
Broadsands
Elbery Fm
Elberry Cove
Fishcombe Point
Duncannon
STOKE GABRIEL
Stoke Gabriel
Waddeton
Galmpton Warborough
CH
Devon South Coast Path
Cemy
Inn
Ashprington
Stoke Point
Woods
Weir
South Downs
Waddeton Court
Galmpton
Hotel
Coombe
Bow
Frogmore
Sandridge
Manor Fm
Quarry (dis)
CHURSTON STA
dismtd rly
Tuckenhay
Bow Creek
White Rock
Long Stream
Middle Back
Pighole Point
R I V E R   D A R T
Galmpton Creek
Churston Ferrers
Corkscrew Hill
Whitestone Fm
Blackness Rock
Lower Gurrow Point
Flat Owers
Brim Hill
Alston Fm
Resr
Cornworthy
Gatehouse
Longlands Fm
East Cornworthy
Dittisham Court
Lower Greenway
Lupton Ho (Sch)
Coomery
Broadgates
Coombe (Hotel)
Dittisham
Ferry
Greenway Ho
Higher Greenway
Lupton Park
Southills Fm
Cott Fm
Anchor Stone
Maypool
MS
Lower Tideford
Bozomzeal Cross
Tunnel
Gitcombe
Kingston
Torbay & Dartmouth Rly
Guzzle Down
Higher Tideford
Broadridge
Fire Beacon Hill
Hillhead Fm
Southdown Fm
Woollcombe
Bozomzeal
Hillhead
Allaleigh
DITTISHAM
Lapthorne
Higher Noss Point
Fort
MS
Higher Dinnicombe
Capton
Downton
Hole
Lower Noss Point
Croftland
Woodhuish Fm
Bruckton
Chipton
Rough Hole Point
Hoodown Fm
KINGSWEAR
Drayton
Shearstone
Stone Fm
Old Mill Creek
Pier
BRITANNIA HALT
Boohay
Nethway Ho
East Hartley
Hemborough Post
Hemborough
Old Mill Cott
Britannia RN College
Dartmouth Harbour
Waterhead Brake
Oldstone
Wadstray Ho
DARTMOUTH
Lower Norton
Ferry
Kingston
Lower Wadstray
Woodbury Fm
Norton
Pontoon
Ferry
Kingswear
Coleton Fm
Blackawton
Hillfield
Bugford Fm
Woodbury
Castle
Ferry
Coleton Fishacre
Cotterbury
Milton
Cotton
Warfleet
One Gun Point
Castle
Higher Brownstone Fm
Waterslade
Broomhill
Wheatland
Gallant's Bower
Castle
Warren Ho
Tower
Coleton Cove
Pruston Barton
Sweetstone
Greenswood
STOKE FLEMING
Swannaton
Worden
Mill Bay Cove
Pudcombe Cove
Dallacombe
Westdown
Eastdown
Ash
Venn
Thorn
Lower Week
Blackstone Point
Newfoundland Cove
Kelly's Cove
Outer Froward Point
Bow
Poundhouse
MS
Compass Cove
Inner Froward Point
Mew Stone

Sailors in small boats have for centuries known that the Dart offers the best refuge from the prevailing westerly winds between Portland Bill and Plymouth, in a harbour with no sand-bar, deep water at any state of the tide, and its high surrounding hills ensuring relatively calm water once inside. It is this factor which led to the rapid growth of Dartmouth from the Middle Ages onwards, though the steepness of these hills made road connections always difficult, and the expansion of the town nearly impossible. By the end of the eighteenth century, however, the largest sailing ships of the day found the Dart too constricting. Small ships of up to three hundred tons were still built there, but the great commercial expansion of this period passed the area by. A renewed period of activity came with the coal-bunkering trade, when, at its peak in 1890, more than seven hundred ships a year called to pick up coal. This trade finally ended in the 1960s and even shipbuilding ceased after 1974.

Commercial traffic in the Dart is now on a greatly reduced scale. Baltic timber ships carry their cargoes up on the high tide to Totnes, perhaps twenty times a year, their red hulls providing a fine splash of colour. There is also an active crabbing fleet, making this the premier shellfish port in the country. Some of their catch goes directly into boats for the Channel Islands and France, the rest by lorry from the wharf at Kingswear.

Dartmouth is however unique in having within it the Britannia Royal Naval College, whose building dominates the harbour, and which trains future naval officers in seamanship on the waters of the Dart. A variety of naval ships, such as minesweepers, patrol boats, frigates and submarines, visit the port for their benefit. There are also Royal Navy helicopters which use the heliport above the town. The existence of the College provides work for the town, improves its social life, and gives visitors much of interest to watch.

People who choose to live beside the Dart, or who come regularly for holidays, are mainly attracted by one or more of three reasons: its great natural beauty and wildlife; its boating or fishing facilities; the peaceful, historic character of its towns and nearby villages. How should a newcomer best sample these pleasures? There is little doubt that the ways used by our ancestors are still the best – on foot or by small boat, the latter being not only preferable but the only way of seeing most of the estuary. There are, mercifully, no motor roads along its banks, and few footpaths.

The best footpaths are at or near the entrance to the harbour, where the National Trust has recently acquired some of the finest stretches of coastline and river frontage on both sides of the Dart, and has opened them up to the public with excellent footpaths. These include a superb coastal walk from just outside Brixham to

Looking north across Dartmouth; on the left is the Britannia Royal Naval College

Kingswear, and another north from Kingswear through Hoodown along the river-side, soon, it is hoped to be extended through Long Wood to Maypool. On the Dartmouth side the public road along the Embankment and Southtown to the Castle is a favourite stroll. The hills above – Dyers Hill and Gallant's Bower – offer good footpaths and wonderful views of the river mouth and open sea. There is also a walk on the edge of the sea outside the harbour round to Compass Cove. A walk from Dartmouth north via Old Mill Creek to Dittisham gives magnificent views over the middle reaches of the Dart. A booklet, *Exploring from Dartmouth*, published by the Dartmouth and Kingswear Society, gives full details of these and many other walks.

For those with only a few hours to spare, the tourist river boats offer a worthwhile trip between Dartmouth and Totnes during the summer. To enjoy the estuary to the full, it is better to spend longer and to go in a small motor boat (which can be hired by the hour) with the tide beginning to flood. First, go outside the harbour and come in as sailors have done since the earliest times. The Daymark (5) above Brownstone on the Kingswear side guides the modern sailor towards the mouth of the river, which is invisible from a few miles out at sea. In the past, the Mewstone and Western Blackstone, now covered with nesting shags and cormorants, were the sailor's only markers. A glance at the headlands on the Kingswear side shows evidence, from raised beaches and cliffs above the present waterline, that once the sea-level was higher than it is now. Underwater, similar ledges and cliffs confirm how the level has varied in geological times as the ice ages came and went.

The harbour entrance is guarded at its narrowest point by the two castles of Dartmouth (6) and Kingswear (7), built between 1388 and 1491 to keep out the French during the Hundred Years War. A chain

Dartmouth Castle at the mouth of the estuary

was stretched from Dartmouth castle in times of danger to a small blockhouse at Godmerock (or Gommerock) opposite. A similar device – a boom from which steel nets were suspended – was used as a protection against German submarines during the last war. It would have been well worth their while to creep in there during the spring of 1944, when nearly five hundred American landing craft were being prepared for the D-Day invasions – the most recent of the long line of ships of war to set off from the port.

Dartmouth was already known as a harbour when it was the assembly point for the European fleet of one hundred and sixty-four vessels about to leave on the Second Crusade in 1147. It sent thirty-seven ships to join Richard I on the Third Crusade in 1190. A flourishing trade with Bordeaux, for a time owned by the English King, grew up during the thirteenth and fourteenth centuries, exchanging Devon cloth and dried fish for wine. The merchants who grew wealthy from this trade built for themselves fine houses, some of which can still be seen in Dartmouth. They also founded St Saviour's church (8), consecrated in 1372, for the growing waterside town. In its chancel lies the brass of John Hawley, generally believed to be the inspiration for Chaucer's 'Shipman' in *The Canterbury Tales*. Famous in his day as a wine merchant, and many times mayor, he led several naval expeditions against the French and was responsible for building the first castle in 1388 – 9.

After the Bordeaux wine trade ended, one with Newfoundland replaced it as the town's main livelihood from the late sixteenth century to around 1815. At its peak, hundreds of ships left annually laden with Devon-made clothes, footwear, iron goods and fishing gear, to sell to the fishermen of Newfoundland in exchange for salted cod. This was taken to the Mediterranean and Portugal and sold to buy dried fruit and wine, which were carried back to England. All the ships used – not large, but sturdy schooners – were built on the Dart, although already by then timber had to be imported from Scandinavia to do so. Reminders of these merchants can be seen in their fine houses on the Quay and in the Butterwalk.

Going upstream, just past the Lower Ferry with its unusual side-by-side tow, is Kingswear Wharf and Railway Station (9) on the east. Opened in 1864 for broad-gauge trains, it was connected to Dartmouth by a passenger ferry. Now the steam train to Paignton is a tourist attraction run by the Dart Valley Railway Company.

The Embankment which runs along the west bank of the river between the two car ferries is an example of how much of the town has been reclaimed from the mud of the river, the original bank of which was at least one hundred yards to the west. The Boatfloat (10), entered today under the bridge on the Embankment, was before 1885 open to the river, and schooners could sail up to the Quay. The Royal Avenue Gardens, and Coronation Park to the north, are both on what were once mud-flats. Under the latter lies a submarine, brought in and scrapped in the 1920s, and buried when the Park was landscaped in the 1930s.

North of the Park is the Higher Ferry or 'floating bridge', operated by paddle wheels and chains. It was started in 1831 by the Seale

View over Dartmouth to Kingswear, with the ferry midway

family to provide the first vehicular crossing of the river. The same family also started the Sandquay shipyards, immediately past the ferry on the west bank. From the 1780s to the 1960s these were the main shipyards of the town, employing hundreds of men. Now they have been converted into a marina.

All the way from the mouth of the river there is a good view of the Naval College on the hill to the north of Dartmouth. Immediately north of the present marina in Sandquay is the area where the college boats are kept, ranging from small sailing and picket boats to patrol boats. There are occasionally larger visiting craft, such as submarines or frigates which may tie up there or moor in mid-river.

The next part of the river – Old Mill Creek – is peaceful. Noss Works, on the Kingswear side, now a marina was originally a shipbuilding yard built in 1893. Taken over by Philip's of Sandquay in 1919, it was in use throughout the Second World War, when it built more than two hundred vessels, including corvettes and mine-sweepers for the Royal Navy. It was badly bombed in 1942 when twenty of its workers were killed.

Jenkins Quay on the Avon estuary

Yealm estuary from Newton Ferrers

Approaching Maypool, on the east bank, with its attractive boat-house beside a small dammed-up creek, look out for some concrete ramps on the water's edge. These were hards built by United States forces in 1944 for their landing craft, which were hidden under trees beside this bank and in many other places along the river. More pleasantly, after Maypool, look up among a clump of Scots pines for the heronry of ten or twelve nests. The birds can also often be seen fishing in shallow water or on mud-banks on both sides. From mid-stream, looking north, one can catch a glimpse of a white house in the trees; this is Greenway House, once owned by the author, the late Dame Agatha Christie. An earlier house on the site was owned by the Gilbert family, and it was there that the brothers John, Humphrey and Adrian were born in Elizabethan times, all to become famous seafarers. Humphrey, knighted by the Queen, searched for the north-west passage and claimed Newfoundland for England before being lost at sea in 1583. Their half-brother Walter Raleigh also stayed in the house as a boy.

On the banks opposite the Anchor Stone, marked as a hazard to shipping, kingfishers can be seen by those willing to wait quietly, and a seal often fishes from the mud on the west bank. The village of Dittisham lies just north, linked by its small passenger ferry to Greenway opposite. Its main street rises very steeply from alongside the Ferryboat Inn. Here the river widens out, and with both outlets invisible, it looks at high tide almost like a lake. It is wise to take no chances: the water is shallow, and the channel marked by buoys round the outer edge of the curve should be followed. Soon, on the east, is Galmpton Creek, another former shipyard now silent, where once Brixham trawlers were built. Bird-watchers should look at the mud-flats to the west of the channel for curlews and oystercatchers.

Beyond Galmpton Creek is an old quarry and tiny quay at Waddeton. These used to belong to the monks of Torre Abbey, who exported the stone by river.

The river bends in a curve to the west, below Sandridge House, where another famous Elizabethan explorer, John Davis, was born, though in an earlier house than the present one which was built in 1805 by John Nash. Davis also looked for the north-west passage, gave his name to Davis Strait to the west of Greenland, and acted as pilot to the Dutch expedition to the East Indies in 1601.

Opposite, the mud-flats are a favourite place for curlews, redshanks, oystercatchers, herons and cormorants. A little further on past Dittisham Mill Creek are some more mud-flats below Whitestone Farm, where dunlin, ringed plovers, curlews, godwits and lapwings are likely to be enjoying its rich protein mixture. Grey seals can sometimes be seen here too.

A creek to the east leads to Stoke Gabriel, up a channel which should be avoided at low tide. The mill pool has been dammed here, to form a pleasant lake, and if the tide allows, a visit to the old church with its ancient yew is a pleasant diversion.

As we continue upstream, a long creek opens up on the west; this is Bow Creek, leading to Tuckenhay. This most beautiful creek is, however, only navigable for a short time around high tide. The southern shores are clothed in beech, oak and other deciduous trees, making it a delight especially in spring and autumn. Occasionally shelduck may be seen. If the tide allows, a visit to Tuckenhay Mill, at the head of the small creek leading off Bow Creek, is worthwhile. This once made high-quality paper, used for bank notes, and has now been converted into holiday flats. Continuing up the main Bow Creek, there is a choice of two good inns, where a drink and meal can be enjoyed at the water's edge, and any left-overs can be fed to some plump mallards.

Returning to the main stream, the channel is now narrow and shallow by Duncannon Reach. Here a path from Ashprington comes down to where a ferry used to cross, to join the road to Stoke Gabriel just over the hill. This is the best part of the river for salmon fishing: Ashprington is recorded in Domesday Book as having two fisheries, almost certainly salmon. It is possible, in the season, to see licensed fishermen at work there with their nets.

Sharpham Woods (11) enhance the west bank as the river now takes three U-turns in quick sucession. Here many cormorants roost for the night in trees. Sharpham House – a fine Georgian house built between 1770 and 1820 – can be seen from the river. In September, this stretch of the river is often visited by ospreys on their passage from Scandinavia to North Africa.

It becomes more and more incredible at this point that the two-hundred-and-forty-foot-long timber ships on their way up to Totnes can negotiate these bends, but they do at high tide, sometimes with only a few inches of water below them. After rounding Fleet Mill, the Totnes home reach stretches ahead, almost straight, with an area of saltings on the west. Soon the Baltic Wharf is reached, so-called because this is where the timber stacked on Messrs Reeves yard originated. A specially constructed turning bay by the wharf enables the timber ships to turn round after discharging their cargo.

Tucked away in the street behind the wharf is the Dartmouth Inn, so-called because many people used to wait there for the tide and a boat to take them down to Dartmouth. Among these were the 'hard lads' signing on in the Newfoundland trade to serve their masters two summers and a winter on the codbanks, many never to return.

Vire Island, alongside Reeves yards, shows that the river here is very shallow. While the estuary is tidal up to Totnes weir, for all practical purposes navigation stops at the bridge. This was once the first place where the river could be forded at low tide upstream from Dartmouth, and explains the growth of Totnes at this point. Linked by ancient trackways to similar crossing places over the Exe, Teign and Plym, the town grew to be one of the four largest in Devon, after Exeter, by the time of the Domesday survey in 1086. It already had a wall, the Normans added a castle, and soon the ford was replaced by a bridge. Down the Dart it exported its products, mainly wool and tin. By the time of the Newfoundland trade, Totnes merchants outnumbered those of Dartmouth in both imports and exports. They spent their profits on fine houses for themselves, many of which can be seen by walking up Fore Street and High Street. One has now been made into a Museum, which presents a good opportunity to see a typical inside layout. An arch across the road marks the place where once the East Gate stood; from there a rampart walk along the top of the town walls leads to the old Guildhall and St Mary's Church, both once part of a priory.

Once this was the fifteenth richest town in England and it now has more listed buildings per head of its population than any other. Clearly, a discerning traveller will enjoy a visit here, as did Daniel Defoe in the early eighteenth century, when he stayed at the Seven Stars Inn, by the bridge. He describes how the landlord netted thirty-three salmon peal (young ones), with the aid of a well-trained dog, in a few minutes, and 'of these we took six for our dinner, for which they asked a shilling, viz. two pence each.' He thought it cheap even then. Sustained by whatever can be offered by the present-day innkeepers – many of whom give excellent value – it is time to return downstream on the falling tide.

This chapter was written by Mrs Ray Freeman on behalf of the Dartmouth and Kingswear Society. Formed in 1959, this is an amenity society registered with the Civic Trust. Since its beginning it has been able to help in the preservation of the beautiful and historic town of Dartmouth and its neighbouring villages, and also promote improvements in a number of ways. The Society is equally concerned with the rural scene, the protection of the coastline, and the preservation of trees. It has more than five hundred members and can be contacted c/o The Harbour Bookshop, Fairfax Place, Dartmouth TQ6 9AB.

Totnes Bridge from Vire Island

East Gate, Totnes

6

Kingsbridge Estuary

The Kingsbridge estuary is the most southerly in Devon. Despite its name and the fact that the lower reaches are sometimes called the Salcombe River, this estuary is, like that of the Dart, more accurately a ria or drowned valley created by Ice Age glaciation and the subsequent rise in water-levels. There is no true river, although small streams flow into the heads of the various creeks which branch out to east and west of the main body of the estuary. It is four miles in length and contains some two thousand acres of tidal water.

The southerly position and the protection from the prevailing winds provided by the cliffs ranging westward from Bolt Head give a remarkably mild climate to the estuary and its immediate hinterland and sub-tropical species of shrubs and trees are not uncommon. The same cliffs shelter the narrow mouth of the estuary from the sea and additional protection is afforded by a sand-bar and the Blackstone rocks (1).

Approximately half the width of the estuary from the Blackstone to Wareham Point is a deep-water channel making Salcombe Harbour a popular haven for yachts and small boats. The creeks and higher reaches dry to mud-flats at low water providing a habitat for many species of wildlife. Sandy beaches extend over the greater part of the East Portlemouth foreshore and at North Sands (2) and South Sands on the Salcombe side.

Apart from the towns of Salcombe and Kingsbridge and the lower village of East Portlemouth, the surrounding area is almost entirely farm and woodland rolling right down to the shore. Agriculture is varied with mixed crops and livestock and the resultant patchwork forms a delightful natural setting for the estuary. Away from the towns public access points to the shore are very limited and the greater part of the estuary can be explored effectively only by boat.

Historical development

The barely discernible remains of earthworks on Rickham Common indicate some prehistoric habitation, but it is fairly safe to assume that subsequently the area was virtually unpopulated forest land until Saxon times.

East Portlemouth and Batson were both mentioned in the Domesday Book and by the Middle Ages the estuary and surrounding arable land supported quite a significant population. Being at a focal point in such a rural area and being accessible at high water to coastal shipping, the development of Kingsbridge at the head of the estuary as a market and trading town was quite natural. The tin industry and expanding cloth trade contributed to the prosperity of the town.

At this time East Portlemouth was the dominant village at the southern end of the estuary with an economy based on agriculture, fishing and shipbuilding. In 1346, for example, the village provided

Aerial view of the upper Kingsbridge estuary, with Kingsbridge in the distance

KINGSBRIDGE
Hospl
Redford
Mus
TH
Schs
KINGSBRIDGE
Dodbrooke
Ranscombe
Marepark Cross
Cross
SHERFORD
Sherford
Bearscombe
Bowringsleigh
Huxton Cross
Westville
Bowcombe 107
Bowden
Sherford Down
Homefield
Resr
113
Upton
Resr 96
58
98
MS
Heddeswell
Sch
West Alvington
Ticketwood
High Ho
Duncombe Cross
43
Keynedon Barton
Coombe Park
Chillington
Hotel
Inn
Preston
Southville
103
103
East Charleton
Frogmore
MS
P
South Milton
87
Easton
Park Fm
Cemy
Inn
The Grange
Frogmore Inn P
Mill Fm
Sutton
65
Auton
WEST ALVINGTON
27
Wks
West Charleton
MS
P
43
27
37
Loo Cross
Oldaway Fm
Youngcombe
Collapit Br
CHARLETON
59
Winslade
Molescombe
82
86
97
Davey Park Fm
99
Gerston Fm
73
Cholwells
Woolston
65
Frogmore Creek
Ham Point
North Pool
68
Kernborough
Bagton
Gerston Point
33
Burleigh Fm
MS
Alston Fm
Blanksmill Br
KINGSBRIDGE ESTUARY
Wareham Point
Halwell Fm
SOUTH POOL
87
Withymore Fm
99
Ilton Castle Fm
Lincombe
Lower Barn
South Pool
Dunstone
Yarde
Ilton
5
Combe
PH
Ford
Tosnos Point
Scoble
Cousin's Cross
115
Inn
MS
81
Shapes Manor
Westerncombe
Gullet Fm
116
alborough
P
Batson
Wilton
74
Kella
Collaton
Cemy
Southpool Creek
Portlemouth Barton
Sch
83
LB Sta
Ford
Goodshelter
Ford
Chivelstone
101
88
38
SALCOMBE
Ferry
East Portlemouth
Waterhead
South Allington
48
Rew
Combe
Harbour
Mill Bay
EAST PORTLEMOUTH
CHIVELSTONE
133
Tor
Hotel
Soar
Castle
South Sands
Rickham Common
Holset
Rickham
West Prawle
134
Borough
Devon South Coast Path
132
East Soar Fm
130
Mus
NT Stink Cove
Sharpitor
132
Portlemouth Down
83
Gara Rock
Hotel
Moor Fm
138
138
Woodcombe
Wr Twr
131
The Warren
Sharp Tor
The Bar
The Bull
Deckler's Cliff
East Prawle
Inn
CG Lookout (dis)
Off Cove
128
Starehole Bay
Shag Rock
Pig's Nose
115
137
Gorah Run
Gorah Rocks
Mew Stone
Ham Stone
Sharpers Head
BOLT HEAD
Little Mew Stone
Ball Rock
Devon South Coast Path
75
Gammon Head
CG Sta
Brimpool Rocks
Langerstone Point
PRAWLE POINT

five ships for the Crecy and Calais campaigns of the Hundred Years War. Piracy, wreck plundering and smuggling were almost certainly profitable sidelines for the local inhabitants, with the last continuing well into the nineteenth century and perhaps beyond.

The decline in the relative importance of East Portlemouth, with a corresponding development of Salcombe, took place in the early nineteenth century when free trade allowed cheap imported food and small farms did not pay. It must have been a serious blow to local morale when the administrator of the Portlemouth Estate, the Duke of Cleveland, caused the two village inns (and, it is believed, a number of the houses as well) to be demolished as the Duchess claimed them to be smugglers' dens. To this day East Portlemouth is a parish with no village inn: are there any others?

From the middle of that century the expansion of small-ship building for the fruit trade created rapid growth in Salcombe. The renowned Salcombe schooners were built all along the waterfront, and indeed at Kingsbridge, and they were amongst the fastest of the vessels in this trade. Speed was essential as fruit loaded in the West Indies, Azores and Mediterranean and brought to major ports such as London, Liverpool and Southampton had to arrive in good condition. Some remarkable voyages are recorded, including a return voyage from the Azores to Liverpool at an average speed of ten knots and a round trip from London to the Azores and back in seventeen days.

A number of prosperous shipowners also lived in and around Salcombe at this time and the town provided skippers and crews for many of their vessels. Some of these in their turn retired to build large houses on the hill overlooking the estuary.

The shipbuilding industry provided a valuable outlet for the Lidstone foundry of Kingsbridge. In addition to ship fittings, the foundry catered for the agricultural and domestic markets. Best remembered locally for its kitchen ranges, it also produced decorative ironwork and many examples of iron railings and fences survived the last war and are still to be seen in the area.

With the advent of steam-ships and the decline in the need for small fast sailing-ships, shipbuilding activities in Salcombe contracted, the emphasis changed to the building of smaller boats and shellfishing assumed a greater importance.

Since the Second World War Salcombe's popularity as a retirement area and tourist centre has grown steadily and to a lesser extent the same is true of Kingsbridge, although the latter has still managed to retain some of the atmosphere of a small market town. Many of the houses in the area are now holiday homes and the resident population has dwindled.

Most of the working boatyards along the Salcombe waterfront have been converted or pulled down to provide waterside houses, many of which remain empty for nine months in the year. Tourism is now the major local industry, although Salcombe is still one of the primary shellfish landing ports in the country.

As part of the Water of Dartmouth in the ownership of the Duchy of Cornwall, the greater part of the Kingsbridge Estuary is leased to the South Hams District Council and is administered by them through the Salcombe Harbour Authority. The estuary is a designated recreational area and, with its hinterland, is within an Area of Outstanding Natural Beauty and a Coastal Preservation Area. At the time of writing, the Nature Conservancy Council is also taking steps to list several areas within the estuary as Sites of Special Scientific Interest in an attempt to protect some of the rare communities and species of marine life which exist there.

The main estuary

As stated earlier, the beauty of the whole estuary can best be appreciated from the water, so let us now take an imaginary boat trip starting from the seaward end.

Towering above us on our left as we head north into the estuary are the two jagged outcroppings of Bolt Head and Sharp Tor sheltering between them the relatively calm waters of Starehole Bay. Lying just below the surface in this bay are the remains of the *Herzogin Cecilie*, a four-masted barque wrecked in 1936 and now a favourite with sub-aqua clubs.

Ahead of us to the right the sweep of cliffs to Prawle Point is topped by downland and fields with the beaches, hotel and old coastguard look-out at Gara standing out conspicuously.

As we reach the Bar the cliffs on our left change to steep-sloping woodland with a few scattered houses. It was here on the Bar that, in 1916, the Salcombe lifeboat *William and Emma* capsized when returning from giving service, with the loss of thirteen of her crew of fifteen men.

Ahead of us the southern slopes of Salcombe town now become more distinct, with large houses standing among broad-leaved woodland. On our left the bay of South Sands is opening up. Look for the distinctive shape of the old lifeboat house at the head of the beach; this was built in 1877 and used until 1925. On the promontory beyond South Sands (3) stands The Moult, a magnificent residence built in 1764 where Tennyson once stayed and, it is said, gained the inspiration to write 'Crossing the Bar'. Then comes North Sands where, tucked into the foot of Moult Hill, stands Cable Cottage, the building which in 1871 housed the English end of the French Atlantic telegraph cable.

As the Blackstone passes astern you can see more clearly the ruins of Fort Charles (4) on the rocks to the left. Built in 1643 on the site of a former fortification, the castle with a Royalist garrison commanded by Sir Edmund Fortescue withstood a four-month siege and bombardment by Parliamentary forces camped on Rickham Common across the water. So gallant was their resistance that when finally forced to surrender in May 1646 Sir Edmund Fortescue and his men were allowed to march out fully armed and with colours flying, and disperse to their homes.

Salcombe waterfront (left)

Moult Hill and North Sands, Salcombe (below)

Tamar estuary at Plymouth from Mount Edgcumbe

Mouth of the Yealm estuary (left)

Aerial view of Kingsbridge at the head of the estuary

On the right beyond the Blackstone the sandy beaches of the aptly named Sunny Cove **(5)** can be approached by footpath. (Frequent swells make landing by boat inadvisable.) As the slopes of Salcombe rise on our left, keep looking to the right where the concrete slipway across the broad sands of Mill Bay is the most obvious remaining evidence of the D-Day preparations. Built by the United States forces stationed in Salcombe it was used to haul out landing craft for maintenance and repairs in the months prior to the Normandy landings.

Further on we come to the old stone pier and enter the main mooring area of the harbour. On our left, beyond the rows of small boats at moorings which dry out at low water, is Batson Creek with the delightful hamlet of Batson at its head. As we continue up the fairway and pass the lifeboat at its permanent deep-water station, look back to see a completely different aspect of Salcombe, with the old town and waterfront at the bottom of the hill and the later Victorian houses and the Church rising above them.

Now we round Snapes Point **(6)**, leaving Southpool Creek on our right, and enter The Bag **(7)**. Ahead of us on either side of the fairway are yachts of all types and sizes swinging to their buoys. A little further on and suddenly, perhaps incongruously, we see an old Mersey ferry, the *Egremont*. This is the floating headquarters of the Island Cruising Club and the base for their dinghy-sailing activities. On the shores beyond *Egremont* is an old limekiln, one of many to be seen around the shores of the estuary. These were used for burning lime, primarily for agricultural purposes.

Leaving The Bag and its moorings behind we pass the mouth of Frogmore Creek and enter the area of open water known locally as Widegates **(8)**. At high water Widegates and the two small inlets to the west provide ideal dinghy sailing conditions for beginners in light weather, and exhilarating sailing for the more experienced on windier days.

From Gerston Point onwards it is prudent to follow the marked channel since grounding on a falling tide will leave us for several hours with little to do except study the activities of the many species of birds in their search for food on the extensive mud-flats.

As we reach the outskirts of Kingsbridge the estuary narrows rapidly and the channel then winds through moored boats right into the heart of the town and the landing steps which are our destination.

The creeks

Southpool and Frogmore Creeks differ widely. With its mouth opposite Salcombe town, Southpool Creek is initially bordered on its southern side by a tidal road from East Portlemouth to Southpool and a long trot of moorings runs parallel to this shore. At Gullet Point **(9)** with its heavily wooded plantation, Waterhead Creek branches off to peter out at the ruins of an old water-mill, while

Southpool Creek bends through a narrow section by the picturesque waterside residence of Gullet Farm before continuing to the village of South Pool.

Frogmore Creek, on the other hand, is virtually uninhabited over its full length except for the village of Frogmore at its head, making it an ideal stretch of water for the naturalist to enjoy.

Ferries

A regular passenger-ferry service runs between Salcombe and East Portlemouth and during the summer months ferries also run from Salcombe to South Sands and, at appropriate states of the tide, to Kingsbridge. These last two offer an opportunity to view the greater part of the main estuary from the water.

The beaches

On the Salcombe side, North and South Sands are readily accessible from the town and a number of small coves can be reached on foot by the more agile and energetic visitors.

On the Portlemouth side, the foreshore and beaches are privately owned from the mouth of Waterhead Creek to the Bull at the outer harbour limits. From the road public access points to these beaches are limited and car parking can be a problem. They can, however, be reached easily from Salcombe by boat or ferry. The public are not discouraged from quiet enjoyment of these beaches, merely requested not to use radios or metal detectors and not to light fires or leave litter.

Attractions for the visitor

Boating

For those trailing boats there are slipways at the Quay Car Park in Kingsbridge **(10)** and the Creek Boat Park in Salcombe **(11)**. Many types of boat can be hired in the estuary and the Harbour Office in Salcombe can provide a list of licensed operators. Boat trips around the creeks and for mackerel fishing operate from and advertise at Salcombe pier. It should be noted that an eight-knot speed limit is in force throughout the estuary.

Sailing

The estuary offers a superb choice of dinghy-sailing conditions for helmsmen of all standards. Organized by the Salcombe Yacht Club, dinghy racing is held each week-end from mid-March until the end of the year. Racing within the harbour offers spectator interest to both visitors and locals particularly during open meetings, the very popular Merlin Rocket Week and the Sailing Regatta. The Yacht Club starting-line can be viewed from public gardens below Cliff Road (12). The main local classes are National Solos and Salcombe Yawls and there is a strong cadet fleet of Mirror dinghies. Some Sunday racing is run by the Island Cruising Club from *Egremont*. Both Salcombe Yacht Club and the Island Cruising Club offer facilities for visiting yachtsmen.

For those not owning their own boat, organized sailing holidays are available in the estuary with tuition as required.

Regattas

There are two Regatta weeks each August: the Salcombe Yacht Club Sailing Regatta offers racing for a wide range of dinghy classes; the Salcombe Town Regatta includes sailing, swimming and rowing races, a water carnival and fireworks among its attractions, many of which are designed specifically for younger children.

The Kingsbridge equivalent of the Salcombe Town Regatta is Fair Week which takes place in July.

Art

For the artist the estuary and its surrounding towns, villages, countryside and coastline provide an abundance of subjects from dramatic to idyllic and it is hardly surprising that local art clubs flourish. Regular exhibitions of their work are held in both Kingsbridge and Salcombe and are well worth a visit.

Museums

In Kingsbridge the Cookworthy Museum commemorates the discoverer of Cornish china clay and the originator of English porcelain who was born here in 1705.

The Salcombe Museum contains a fascinating selection of items of local and maritime interest.

High on the bluffs above the Bar, Overbecks Museum features shipbuilding and natural history and contains a secret room for children, with dolls and toys. The adjacent Sharpitor Gardens comprise six acres of rare and exotic plants, shrubs and trees and give spectacular views of the estuary.

Walks

There is an almost limitless choice for the rambler, from the rugged and breathtaking (in both senses!) National Trust coastline to the gentle creekside Sunday afternoon stroll along the level road from Salcombe to Batson (choose high water for this one). In primrose time the old coaching road from Kingsbridge to Salcombe via Collapit and Blanks Mill bridges is a delight, and careful planning will allow the option of making the return journey by ferry.

Wildlife

In the early 1960s, when it became apparent that increasing recreational use of the estuary was bringing pressures to bear on the wildfowl's environment, the Kingsbridge and District Pigeon Shoot Club and the Devon Trust for Nature Conservation negotiated jointly with the Duchy of Cornwall for a licence to manage the foreshore for both conservation and shooting with the future requirements of both activities in mind. As a result, areas for shooting and areas for resting and feeding have been agreed and wardens provide some measure of control where none existed before.

Wildfowl counts are difficult to assess as the population varies considerably depending on the severity of the winter, for example, or seasonal variations in food supplies. However, there is no doubt that resident mallard have increased noticeably in number and it is reasonable to assume that other species have also benefited as a result of this continuing policy.

For the knowledgeable, the many miles of uninhabited shoreline abound with interesting flora and fauna, but even in the busier areas such as The Bag, herons, kingfishers and otters can still be seen.

Sea angling

The Kingsbridge estuary offers considerable opportunities for the angler. Within the estuary the most likely species to be caught by beach casting are dab or flounder. For the more adventurous who are prepared to cast from the rocks good sport can be had from Limebury Point (13), the Tea Garden rocks off South Sands or the Castle Rocks around Fort Charles. Here from early summer until really cold weather sets in during late December/early January, first-class opportunities are offered for fine bass or pollack, although even here excellent flatties can be caught.

For the boat angler fishing at the mouth of the estuary is unrivalled. In suitable weather conditions a small boat can give the angler an excellent day's fishing, covering the whole range from wrasse, bass and pollack to plaice and dab.

Taking in the wider marks of the mouth of the estuary from the Mewstone to Gammon Head one can catch fine specimens in the species so far mentioned and additionally fine ray and conger.

Mallard duck and drake

Live bait is readily available within the estuary, with lugworm, ragworm and mussels being evident at North Sands at low water and around the rocks on either side of these sands. For those more determined to catch bass there is an ample quantity of live send-eels, at appropriate states of the tide, from Sunny Cove right up to Fisherman's Cove (14).

The Salcombe and District Sea Anglers Association organizes competitions and an annual fishing festival in the autumn.

With its climate and the variety of its scenery and attractions it is little wonder that the Kingsbridge estuary holds an irresistible magnetism which draws people back again and again.

This chapter was prepared by the Salcombe and Kingsbridge Estuary Association. The Association was founded to satisfy the need for a single organization to co-ordinate the views of all the various interested parties on matters of mutual concern relating to the estuary and to express them to the authorities which are in final control of the estuary and its environs, thus consolidating the powerful influence of which the users working together are capable. The Association has its own representative with the Harbour Authority. The objectives of the Association are to preserve the essential character of the estuary; to encourage public participation in any decisions which may affect the area and its residents; to develop policies on which planning decisions should be based, and represent them to the appropriate authorities; to prepare specific proposals for improving facilities where necessary and generally to serve the interests of the users of the estuary.

Membership of the Association is open to all persons and organizations concerned with the use or conservation of the estuary and its environs. The Association can be contacted through the Secretary, Rosemary Singer, 'Braemar', Herbert Road, Salcombe, Devon, TQ8 8NH.

7

Avon Estuary

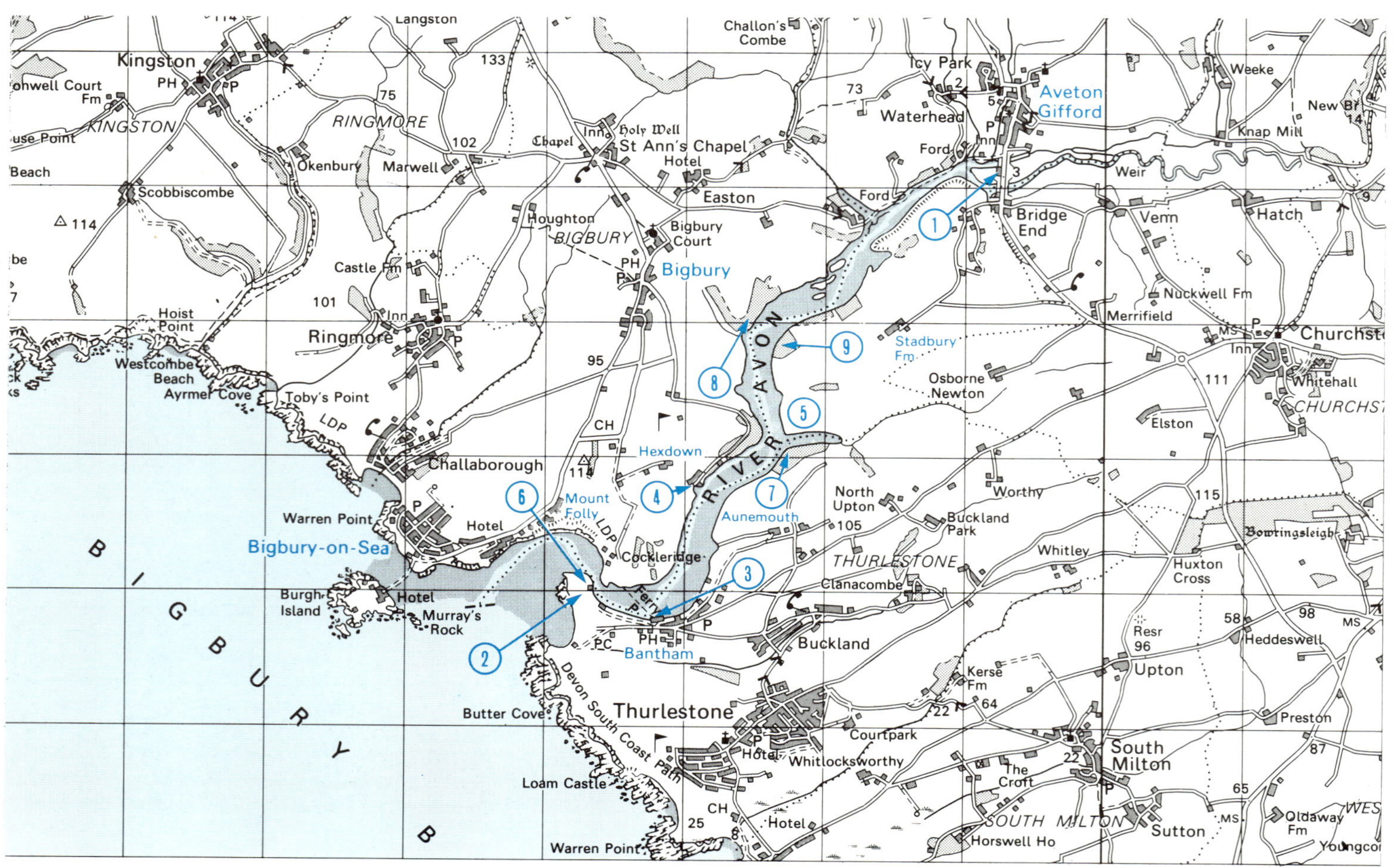

The estuary comprises about four miles of tidal water from the weir above Aveton Gifford Bridge **(1)** down to the sea. The River Avon runs down a wooded valley with, at low tide, mud-flats and sandbanks. The whole area is designated as being an Area of Outstanding Natural Beauty and a Coastal Preservation Area.

The estuary was, in former times, of greater importance than it is today. An ancient pre-Roman camp near the mouth of the river **(2)** was in a strong position, defended on three sides by the sea and the estuary and on the fourth side by a marsh. The site was uncovered by the great gale in 1703 which destroyed Winstanley's Eddystone Lighthouse. The site is now obliterated by blown sand caused by the encroachment of the sea and the disappearance of the marram grass or 'bents' from which the name of Bantham is derived. Cartloads of bones from the camp midden were taken away by the local community to fertilize their fields. Some of the artefacts collected towards the end of last century may be seen at the Torquay Natural History Museum.

Commercial traffic on the estuary was considerable, with Bantham quite a busy little port. Barges and small vessels bringing cargoes of limestone, coal, artificial manure, farm implements and similar items were off-loaded at Bantham Quay **(3)** and then reloaded with corn, potatoes and so forth before sailing away. A pilot who lived in Bantham rowed out to the bar to escort incoming and outgoing vessels. He also ferried passengers across the river from Bantham to Bigbury.

Bantham Sands and Ham at the mouth of the Avon estuary

Bideford Bridge

Torridge estuary at low tide

Three old limekilns border the river, one at Bantham, another at Hexdown Quay **(4)** and another at Stiddicombe Creek **(5)**. At these imported limestone was rendered suitable for agricultural use by burning it from below by a coal fire.

The old thatched boathouse on the salmon quay **(6)** (locally known as 'Jenkins Quay') was formerly a ferry point. Salmon netting was carried on successfully at this deep part of the river until 1981. At Bantham Quay stand the Pilchard Cellars. The building, now converted to a private residence, was in past times used for storing and salting fish when pilchard were plentiful in Bigbury Bay.

A broad flat expanse of sand below Bantham Ham is easy to reach, while across the river steep cliffs rise from a narrow ridge of sand. Further up the river the foreshore becomes more stony and the banks, clothed in oak, hawthorn and blackthorn, rise to about three hundred feet. The rock is slate and sandstone dating from the Lower Devonian Age. There are small woodlands at Stiddicombe **(7)**, where sweet chestnut trees predominate, at Doctor's Wood **(8)** a little further upstream on the Bigbury shore, and, almost directly opposite, Stadbury Wood **(9)**.

A marshy area and muddy creeks in the upper reaches of the estuary between Doctor's Wood and Aveton Gifford provide feeding grounds for wildfowl and waders. Species recorded include oyster-catcher, mute swan, ringed plover, lapwing, dunlin, curlew, red-shank, sandpiper, mallard, teal, shelduck and wigeon. In Stiddicombe Wood is a heronry, and these silent grey birds may be seen at the water's edge at all seasons. In the steep rough ground high above the water, badgers, foxes and other wildlife can be observed by the patient naturalist. Butterflies, moths, lizards and adders can breed undisturbed, as can the small birds such as long-tailed tits, coal tits and goldfinches.

There are few points of access to the water. From the A379 over Aveton Gifford Bridge the tidal road is signposted and may be used by vehicular traffic only at low water. A single-track, unclassified road leads to Bantham village from Bantham Cross on the A379. Foot-paths give walkers good views of the estuary. From Bigbury the path from Folly leads down to the water. From there it is possible to cross to Bantham by the ferry (operating only at fixed times during the summer months) and then to follow the cart-track and footpath from the village, via Aunemouth to Stiddicombe Wood. The path skirts the end of Stiddicombe Creek and climbs the hill above the old lime kiln and on to Stadbury. From here it is not far to the main Kingsbridge to Plymouth road.

For those interested, there is fishing in the estuary for bass, flat-fish and mullet. The sandbanks provide the fisherman with lugworm and sand-eel for bait. Sailing, canoeing, swimming, bird-watching, and picnicking are all popular on this beautiful and unspoilt estuary. An interesting old footway at the end of the sandspit opposite Bantham Quay leads to Hexdown Quay. This path

Jenkins Quay and boat-house

Cormorant and gulls on mud-flat near Aveton Gifford

has been hewn out of the rock. Legend has it that it was used by smugglers, but it was more probably used by coastguards. The path is negotiable at most tides with the exception of high spring tides.

As the seasons change so do the colours of the estuary. Fields, trees, even the water, change from a misty grey to a sunset red and all contribute to the quiet beauty of this unique waterway. There is growing public appreciation that the conservation of Devon's estuaries is of the utmost importance and that every effort should be made to preserve this heritage for the enjoyment of future generations.

This chapter was prepared by Anne Jenkins, A.R.A.D., F.I.S.T.D., and Dr M. D. Dixon on behalf of the Aune Conservation Association. The Association was formed in 1979 with the aim of preserving the essential character and the quiet amenities of the Avon Estuary, and to protect its wildlife. It has 110 members. Enquiries should be addressed to the Secretary, Burnt House, Buckland, Kingsbridge, Devon TQ7 3AF. The Association uses the old name in its title in order to preserve its use and to distinguish the Devon river from other Avon rivers.

8

Erme Estuary

The 'baby' Erme is the smallest of the South Hams rivers, but the charm of its estuary is perhaps the greatest owing to its virtually unspoilt nature. No new building has taken place since the last century and the whole, besides being in an Area of Outstanding Natural Beauty, is now a Site of Special Scientific Interest.

The Erme estuary has always been a difficult one for sailors because of the East and West Merries (1), two lines of rocks guarding the mouth of the river, which are exposed at low tide and extend from the eastern cliff to within a stone's throw of the western cliff, where alone a deep channel with a sandy bottom is available for small vessels desiring to enter the river. It was possibly on these rocks that Philip of Austria and Juana of Castile were wrecked in 1506 on their way from Antwerp to claim the Kingdom of Castile. Tristram Risdon in his *Continuation of the Survey of Devonshire*, published in 1714, says:

The next river that emptieth itself into the sea is Arme whereof the place is called Armouth, which is so shallow and full of Rocks, that no ships arrive there without Extremity of Weather inforce them. In the Reign of King Henry VII, Philip King of Castile landed there by a tempestuous storm with the loss of two of his Ships.

John Britton and Edward Bragley in their *Beauties of England and Wales*, published in 1803, describe 'a singular phenomenon' that happened at the mouth of Mothecombe Harbour in 1798.

From the cliff on its western side projects a peninsula of many acres called 'Mothecombe-Back', consisting of an accumulation of sand and gravel, which has resisted the force of the waters time immemorially, and has a fair annually held there on it. This peninsula reaches so nearly across the harbour that the River Erme is confined by it almost close to the eastern cliff and there flows into the sea. At the beginning of the above year, in a tempestuous night, the waves formed another back, or peninsula, across

the harbour, almost as large, and apparently as firm as the ancient one: this seemed joined to the eastern cliff, as firmly as the other to the western cliff, and in consequence, the river, after clearing the old back, was forced to run quite across the harbour by the side of the new obstruction, before its waters could unite with the ocean. This occasioned so great an impediment to the navigation of the river (which was chiefly by vessels in the coal and culm trade) that meetings were held to consult on the possibility of regaining the passage, by cutting through the new back. Before, however, determination was made, and after the formidable barrier had continued undiminished for several months, the sea, in another stormy night, washed the whole away, leaving the harbour in its former state, the ancient peninsula not being in the least affected.

The ancient peninsula has now entirely disappeared, though perhaps not so very long ago, as Mrs Bingham Mildmay, born in 1835, remembered a large expanse of grass in front of the existing Coastguards' Cottages.

Coastguard cottages at the mouth of the Erme estuary

SC
Southwood Wood
Choakford
Swainstone
Brook
Wks
West Pitton
Popple's Br
Coyton
Fursdon
West Worthele
Marjery Cross
Caton
East Worthele
Weir
Keaton
Penquit
Manor Ho
Ludbrook
Dunwell
ERMINGTON
Tod Moor
Thornham
East Pitton
Lyneham Ho
Lotherton Br
143
Preston
145
Shils Bar
Higher Ludbrook
MS
Strode
Treby Fm
138
Westlake
Ludbrook Manor Ho
Winsor
Airstrip
Inn
Wks
Weeke
YEALMPTON
Worston
Luson
Ermington
Shilston Br
Yeo Fm
Burraton
Hotel
MS
Fawns
Sheepham
Stoneycross Fm
Longbrook
51
Wilburton Fm
Ermington Wood
Orchard Fm
Clickland
Hollowcombe Fm
18
Mary Cross
Yeo Park
Long Brook
Sexton
Edmeston
MS
Stok
MODBURY
Way Fm
MS
Sexton
Fancy
Yealmbridge
Sawmill
MS
Sequer's Br
16
MS
Modbury
Shire Horse Centre
Dunstone
Flete
Weir
Goutsford Br
1643
Torr
Ramsland
Hole Fm
Little Orcheton
Stoliford
Crebar
Ford
Little Modbury
MS
Holbeton
R ERME
Ashridge
Whympston
Combe
Luson
PH
P
Great Orcheton Fm
Gnaton Hall
HOLBETON
Efford Ho
Oldaport Fort
Shearlangstone
Creacombe Fm
Borough Fm
Whitemoor
Tor Rock
Seven Stones Cross
Cumery
Brownstone
Clyng Mill
Langston
Combe Fm
Preston Fm
Coombe Fm
Torr Down
Great Torr
South Langston
Tuffland
Alston Hall (Hotel)
Battisborough Cross
Haye Fm
Pamflete Ho
Kingston
Challa Cor
Poole Fm
Keaton
Mothecombe
Wonwell Court Fm
PH
P
RINGMORE
Carswell Fm
Owen's Hill
KINGSTON
Chapel
Inn
Holy Well
Lambside
St Anchorite's Rock
Gull Cove
Battisborough Sch
Malthouse Point
Okenbury
Marwell
St Ann's Chapel
Hotel
Blackaterry Point
Bugle Hole
Erme Mouth
Wonwell Beach
Scobbiscombe
Easton
Wadham Rocks
Butcher's Cove
Battisborough Island
Houghton
Bigbury Court
y Island
Mary's Rocks
Fernycombe Beach
BIGBURY
Bigbury
Beacon Point
Castle Fm
PH
Hoist Point
Inn
Meddrick Rocks
Westcombe Beach
Ringmore
Ayrmer Cove
Toby's Point
LDP
CH
Hexdown
Challaborough
Warren Point
Hotel
Mount Folly
R I V
Aner

Shelduck prospecting for nest sites

Today the estuary is wide and shallow, completely covered only at high tide. Because of this, passing yachts rarely visit and moorings, which dry out, are limited to twenty-five.

The estuary was previously much more used and navigable than it is today. An Iron Age fort was established at Oldaport (2) and later the woods bordering the river were coppiced for charcoal for burning in the limekilns (3), the produce of which was taken away by barge. Coal was carried up the river and grain down it, mostly for and from the villages of Holbeton, Ermington and Kingston. In the last fifty years the increased drainage of the agricultural lands bordering the higher stretches of the river has meant that the Erme has become more and more of a spate river in the winter, rising and falling in hours rather than days. This increased speed and volume of water has silted up the estuary to such a degree that the ford between Efford and Saltagrass (4) is now no longer passable even at low tide. It was by this ford that the people of Kingston went to Holbeton and on to Plymouth, avoiding the long detour via Sequers or Sackers Bridge on what is now the A379.

The wildlife of the estuary is varied and constantly changing, with kingfishers, herons and shelduck breeding each year. Migrants include wigeon, teal, goldeneye and tufted duck, sometimes in great numbers. Buzzards can nearly always be seen soaring over the woods and ravens scavenge from the cliffs. The flora and fauna is typical of the south-coast salt-marshes, comprehensive in its content, but without any great rarities. Sea trout and a few salmon run up the river, together with mullet and bass. Above all it is the peace and the feeling of a quietly unchanging existence which is the main attraction of this small estuary.

A unique feature of the Erme estuary is that all the lands bordering it have been under one ownership for more than one hundred years, enabling the unspoilt nature of the estuary to be preserved. Public access to Mothecombe Beach is available on Wednesdays, Saturdays and Sundays, with car parking in a small car park (5) at the end of Mothecombe village. Access can be obtained to Coastguards' Beach (6) and Wonwell Beach via the council roads, but the parking at Wonwell is very limited, and there is no parking on the slipways. These are the only two points of public access to the estuary besides the Coast Path. It is, however, possible for groups or schools to obtain permission to walk along the old carriage drives which run alongside the estuary by writing to The Flete Estate Office, Mothecombe, Holbeton, Plymouth, Devon PL8 1LA. Numbers are controlled so as to minimize disturbance to the habitat and wildlife.

This chapter was prepared by Mr A. J. B. Mildmay-White on behalf of the Flete Estate. The Estate owns the whole of the estuary and adjacent lands and manages the area so to preserve its natural beauty and peaceful character. The Estate Offices are at Mothecombe, Holbeton, Plymouth, Devon PL8 1LA.

9
Yealm Estuary

Newton Creek from Noss Mayo

The River Yealm rises on Dartmoor and flows to the South Devon coast. The estuary extends from Puslinch and Brixton, near Yealmpton, southwards for about two and a half miles to Newton Ferrers, and thence westwards for half a mile to enter the sea at the eastern end of Wembury Bay. It has long been designated officially as an Area of Outstanding Natural Beauty, its waters lying embedded in a richly wooded setting where only in the limited built-up village areas is there any road access to its waterfront.

There are some fourteen miles of tidal shoreline, about six of which are backed by woodland, about two by built-up areas, and the remainder either by farmland or cliffs and uncultivated slopes. Geologically, its setting is in Old Red Sandstone, meeting at Puslinch Devon Limestone in which underground water has hollowed out the Kitley Caves and volcanic cooking has produced the fine Kitley marble. The villages of Newton Ferrers and Noss Mayo are separated by the mile-long Newton Creek which joins the main estuary at the harbour anchorage known as Yealm Pool. Higher up the estuary Cofflete Creek, of similar length, stretches from Brixton Torr (1) to Steer Point. Still further up, the Silverbridge valley entering Mudbank Creek has been dammed to form Kitley Pond (2), an ornamental freshwater lake in front of Kitley House and a valuable adjunct to the estuary's resources of food and shelter for wildlife.

Yealm Pool – that part of the main estuary from just above Misery Point (3) as far as Madge Point (4) – provides a most attractive anchorage for yachts and small craft. At the mouth of the estuary a bar of clear sand, exposed only at low spring tides, reaches across from Season Point, narrowing the navigable channel to the southern side. A sand-spit marked by a buoy extends into Yealm Pool from near Warren Point. This beautiful estuary is popular with visiting yachtsmen, but care is necessary, and without local knowledge, navigational charts are essential.

The church at Stoke – St Peter the Poor Fisherman **(5)** – was the daughter church of Yealmpton and the inhabitants had to carry their dead to be buried there. The churchwardens of Yealmpton were responsible for ensuring that the beacon at Worswell was kept in good order.

The village of Noss Mayo was not mentioned until 1309 when it was recorded as being one of the many newly established market towns which had sprung up all over the country in the wake of an expanding economy and population. Not suprisingly, it was not a successful market but, like Newton, it established itself as a self-supporting community and changed little over the centuries.

In the 1870s Edward Charles Baring bought Membland, then a large estate in the parish of Holbeton with farms in Noss Mayo. Over the next ten years he changed the area as he built new farmhouses and houses for his staff and he brought a new prosperity to the village. In 1880 he asked the parishioners whether they would prefer the restoration of their church on the cliffs at Stoke or a new one in the village. They chose the new and he built one to rival Newton's in position and to better Newton's in furnishings. A few years later, Lord Revelstoke, as he had become, sold up and left following the near collapse of Baring Bros, the bankers. His legacy of a distinctive local architecture and a lasting tie between Membland and Noss Mayo remain.

During the last ninety years the separate strands of Newton, Puslinch, Noss Mayo and Membland have come together and now they all form the civil parish of Newton and Noss and the united benefice of Newton Ferrers and Revelstoke. The old rivalries between the villages now appear only during the rowing regattas and the church fairs.

Churches

By the year 1086, a church stood on the site of the Holy Cross, Newton Ferrers **(6)**; it was dedicated to St Mary, as were so many

The church of St Peter the Poor Fisherman at Stoke

churches of that time. Of that church nothing now remains above the ground, although the foundations of that smaller church were uncovered during the restoration work of 1885.

The present church is of thirteenth-century origin or earlier, although many changes have taken place since that time. In 1342 a bishop's commission was complaining that the church at Newton Ferrers had not been rededicated after its enlargement, and we know that the present tower was added in the fifteenth century. Originally the tower had granite pinnacles at its four top corners, each of which was surmounted by a cross; these were removed at a later date to prevent the tower from splitting.

In the past the interior of the church would have appeared differently. Until the eighteenth century there was a wooden screen across the chancel. Parts of this were rediscovered under the floor during the restoration, and these are now incorporated into the screen at the west end of the church. In the days when a church band provided the music at services there was a minstrel's gallery at the west end of the church.

During the Civil War the rector here was Edward Elliot, whose tombstone is fixed to the wall at the west end of the churchyard. He was not ejected from his living as so many were, but at his death in 1644, no rector was appointed until the Restoration. The large stone tombstone in the south aisle is believed to be that of William Bradbridge, who was Bishop of Exeter as well as being Rector of this parish. For several centuries Holy Cross stood watch over its parish alone, and then in 1882 it was joined by the church of St Peter, Revelstoke **(7)**, built by Lord Revelstoke to replace the church of St Peter the Poor Fisherman at Stoke, which had fallen into disrepair.

From at least the early fourteenth century the church of the Poor Fisherman met the spiritual and social needs of the people of its parish, although until the fifteenth century the dead had to be taken the three miles to Yealmpton to be buried. In 1840 the church was badly damaged by a storm and from then on worship took place in a chapel of ease at Noss Mayo (now the village hall).

The church of St Peter, Revelstoke was built all in one period and is a gem of its kind. Lord Revelstoke, its builder, lies buried outside the east end of the church.

The two churches at Newton and Revelstoke with their interesting past stand as living witness to the stability, the care and the unchangeableness of God. They are supported by two lively congregations who strive to continue the work and worship and service that has surrounded the churches over the centuries.

Walks and viewpoints

Facilities in the estuary are limited, car parking is difficult and beaches can be reached conveniently only by boat. Other than the approach by sea, this attractive estuary can most easily be appreciated by using the excellent walks on the National Trust property.

This comprises nearly six hundred acres embracing a variety of beautiful scenery: sheltered woods, breezy cliffs and open farmland. The South Devon Coast Path passes through the area, but a ferry to cross the River Yealm operates only during the summer months. So the National Trust walks are divided by the river into Wembury Cliffs, west of the Yealm, and Yealm Woods and Warren Cliffs, east of the estuary.

Wembury Cliffs

The south-facing cliffs between Wembury Beach and the Yealm Estuary are best approached from the larger National Trust car park at Wembury **(8)**, situated below the church. Here can be found a cafe and a National Trust shop. A path runs eastwards towards the Yealm Estuary and provides excellent views across Wembury Bay to the Great Mew Stone and the mouth of the Yealm.

Beyond the old Rocket House (now a private dwelling) **(9)** take the left fork which runs along the top of Warren Point. The view from this high ground up the River Yealm to Dartmoor, along the Newton Creek, and out to sea, has a variety of interest not often encountered on a coastal walk. Now drop down almost to sea-level, past a landing stage. The path returns past Warren Cottage, providing a view of the wooded slopes, mostly owned by the National Trust, across the anchorage. The outward path is rejoined near the Rocket House, and the return to Wembury provides views across Plymouth Sound to the Cornish coast, with the Royal Navy gunnery school at H.M.S. *Cambridge* in the foreground.

Yealm Woods and Warren Cliffs

The coast is conveniently shaped to provide circular walks of five or seven miles. It is recommended to leave cars at the National Trust Warren car park **(10)** some two hundred yards back from the cliffs. The temptation to use the track from Noss Mayo through Ferry and Passage Woods by car should be avoided. This track is narrow with few passing places and no parking area.

The track round the cliffs was built on the instructions of Lord Revelstoke about a hundred years ago as a carriage drive round his property, the Membland Estate. This is the Revelstoke Drive, or Nine Mile Drive, and the task gave winter work for the fishermen. The gradients are easy and the surface reasonable, except where it passes through woods, where, after rain there are muddy patches.

The National Trust Warren car park provides the best starting point for walks in this area as it gives two alternatives. If a five- or seven-mile walk is desired, then instead of making straight for the cliffs, drop down to Noss Mayo or Ferry Cottage **(11)** and follow the coastal route round in an anti-clockwise direction. Alternatively one may go westwards along the cliff track which provides splendid views. A loop path may be taken to Blackstone Point on Warren

Beach, rejoining the track at Warren Cottage. Proceeding westwards there are views across the open waters of the Channel, and in front will be seen the Great Mew Stone, Rame Head and the Cornish coastline. On a clear day look out for the Eddystone Lighthouse which may be seen on the horizon. One field back from the coast track the Warren Wall can be seen. This massive wall was built to keep rabbits in their place on the cliffs when the area was used for their commercial breeding.

The track continues, with constantly changing views, beyond Gara Point to Mouthstone Point where a splendid view of the Yealm Estuary can be obtained. Cellar Beach **(12)** is approached by a small path leading off the main track. At this point one can either continue round the circular walk or retrace one's steps back to the car park. If the latter route is chosen it provides new panoramic views across Bigbury Bay to Bolt Tail and Bolt Head at the entrance to the Salcombe Estuary.

Bird life

Not only is there diverse bird habitat along the river, the estuary and its environs, but the area also occupies a prime position on both coastal and land migration routes. Birdlife is thus plentiful, varied and of interest at all seasons, and unusual visitors are not infrequent. The keen observer is particularly fortunate in that there are public paths of easy access, from the villages and Puslinch, round the headlands at the estuary entrance and along the river banks, providing easy and unimpaired viewing.

There is always shore and seabird activity round the mouth of the estuary. Large numbers of gulls, cormorants and shags nest on the Mew Stone and the entrance cliffs, and a small colony of fulmars now appears established at Season Point. Gannets are often seen in the bay and terns regularly pause during migration and fish up river.

Cormorant

After storms at sea guillemots, razorbills and, more rarely, little auks and various divers come in for shelter and rest.

The scrub, gorse and heather slopes atop the headlands harbour stonechats, linnets, yellowhammers and blackcaps and many migrants make landfall here from the Continent or gather for departure.

The upper river and estuary broaden into shallow tidal waters and mud-flats from Steer Point to Puslinch, with a small salt-marsh at its head. Here there is a long-established heronry of ten to twelve pairs. Kingfishers are always evident and a few shelduck also breed. Abundant gulls find shelter here in winter with wader and wildfowl populations. All wader species have been recorded on passage. Curlew, oystercatchers and redshank are the most numerous species to stay the winter, but a few greenshank, godwits and common sandpiper may also remain. Influxes of lapwing and snipe occur during colder spells. Rare visitors include the osprey and little egret, while purple and night herons have been recorded. These waters attract only moderate flocks of wildfowl, mainly mallard, teal, wigeon and shelduck. Geese are absent, apart from short visits by wandering Canada geese. Small parties of dabchick are not unusual.

The buzzard is a constant feature soaring in the sky and many pairs breed in surrounding woodland. Kestrel and sparrowhawk are also regular breeders, while peregrine and merlin may be spotted occasionally during their winter excursions.

Ravens are also a feature of the area, nesting along the coastal cliffs, their harsh croaking calls readily identifying them amongst the local hordes of rooks and jackdaws. Of the owls only the tawny is common, the little owl is rarely seen, and, alas, the barn owl is now a rarity. Sizeable flocks of redwing, fieldfare and golden plover often retreat southward in severe winter spells to our milder coastal climes and chiffchaff, blackcap, black redstart and firecrest will occasionally overwinter.

A visit to nearby Wembury Point is always worthwhile. It is one of the few strongholds of the cirl bunting and holds a winter flock of purple sandpiper, turnstone and grey plover.

Marine life

Beneath the water at high tide the Yealm estuary is just as variable in teeming life as any similar area of dry land. At low tide in the side creeks and upper estuary the bare mud-flats conceal myriads of burrowing sandhoppers, ragworms of many kinds and furrow shells. While the iridescent bristly ragworms catch smaller animal life with large black jaws, the furrow shells vacuum the surface with long feeding tubes for fine food particles. All these animals are food for birds at low tide but, as the tide rises, it brings the basic food for all, fine plankton and decaying seaweed, as well as marine predators such as foraging shore crabs and fishes.

The low cliffs bordering the estuary are covered by short black channel wrack, but below there are flatter rock terraces and outcrops covered in larger seaweeds, the saw wrack, bladder wrack, flat wrack and knotted wrack. At high tide the prawns tickle the toes of the paddler and at low tide the wracks afford shelter, food and anchorage for many animals including crabs, winkles, limpets and sandhoppers. Near the mouth the wracks are peppered with tiny white spiral calcareous worm tubes and, between their holdfasts, bright orange or yellow sponges compete with barnacles to coat the rocks. Inseparable from the rocks themselves, the millions of encrusting barnacles open their off-white pyramids only at high tide to fling out their legs and kick food into their mouths. Limpets move slowly in circles to browse their own little rock pastures.

On the beach between and below the rocks all the larger flat stones are adorned with tangled white snake-worm tubes, barnacles, limpets and holdfasts of various wracks.

Below half-tide the muddy gravel is rich in many kinds of worms and at spring low tides sandbanks seldom exposed can be the source of gaper shells, six-inch razor clams, 'pullet' carpet shells the colour of pullets' feathers, triangular trough shells and sunset shells which are yellowish and have purple rays. Here also six-inch burrowing sea cucumbers with pink transparent sticky skin and twelve-branched tentacles at one end tangle their food in sticky threads and meaty ten-inch pearl-grey 'cat' ragworms with red tints and soft yellow bristles hide in the sand.

In the shallows in summer one may be lucky enough to find a six-inch sea hare, rough to the touch and red to olive-green in colour, which can secrete purple slime and has swum into the estuary to lay its long strings of orange or pink spawn. Or, if very lucky, the paddler may find a baby cuttlefish, a fascinating sight with its changing body patterns, acrobatic swimming and puffs of ink.

Near the mouth the channel contains a multiplicity of species of molluscs, worms, hermit crabs, shore crabs, spider crabs and starfish. Here in 1976 'japweed' was found and, although a nuisance to yachtsmen, it has not choked the harbour as predicted and it has provided a stimulus for the introduction, growth and maintenance of yet more forms of animal life. Bass, flounders, plaice, mullet, rays, the Cornish sucker and a few other species of fish penetrate the estuary and now provide sport for anglers or paddling children where less than thirty years ago they provided food for a school of dolphins which regularly entered the mouth.

What other place can give more living interest than a South Devon estuary?

This chapter was prepared by members of the River Yealm and District Association with the assistance of the National Trust. The Association was formed with the aim of preserving the natural beauties of the tidal water of the River Yealm and surrounding countryside and villages. The Association represents more than 500 residents. The Chairman, Major F. W. Garth, may be contacted at Rainam, Yealm Road, Newton Ferrers, Plymouth, Devon PL8 1BN.

10

Tamar Estuary

Introduction

The Tamar estuary is the larger of the two estuaries leading off Plymouth Sound, but in comparison with the Plym, it is little known, unpolluted, beautiful and largely unspoiled, and deserves to remain so. It is bounded to seaward by a line joining Devil's Point and Wilderness Point **(1)** and includes all waters to the northward of that line, as far as the tidal limit.

The River Tamar rises near Morwenstow in Cornwall and flows roughly southwards for about forty-one miles to Weir Head **(2)**, whence it is tidal for the last nineteen miles of its course. In A.D. 936, the river was established by King Athelstan as the boundary between Celtic Cornwall and Saxon England. The River Tavy, which joins the Tamar at a point six and a half miles from the mouth, has its source on Dartmoor, its tidal section is four miles long, with its head at Lopwell Dam. Between the Royal Albert Bridge **(3)** and the mouth, the Tamar Estuary is known as the Hamoaze. At four miles from the mouth, it is joined by the River Lynher, which rises on Bodmin Moor and is tidal for eight and a half miles from Notter Weir. The tidal reaches of these three rivers, plus those of the River Tiddy, which is a tributary of the Lynher, constitute the Tamar estuary system. It is all part of the Dockyard Port of Plymouth, and under the jurisdiction of the Captain of the Port. The only part which lies in Devon, however, is that portion to the east of the medial line of the Tamar; the remainder, including the estuarine branches of the Lynher and Tiddy, lie in Cornwall. This description of the estuary will therefore concentrate on the eastern side of the main branch (the Tamar) and on the Tavy branch.

The Tamar estuary near North Hooe

Buckland Monachorum
BUCKLAND MONACHORUM
Bickham
BICKLEIGH
Venton
Robborough
Broadley
Pound Cross
Porsham
Bame Wood
North Coombe Fm
Ashleigh Barton
Southway
Sch
Whitleigh
Uppaton
Alston
Coppicetown
Inn
Crapstone Ho
Milton Combe
Lopwell
Lopwell Dam
Creekside
Marsow House
Gnatham
Blaxton
Blaxton Wood
Horsham
Tamerton Foliot
Ph
Higher Walreddon
West Down
Rivet
Brooktor
Berra Tor
Pound
Didham Fm
Buckland Abbey
Collytown
Ford
RIVER TAVY
Warleigh Wood
Warleigh Ho
Walreddon
Morwelldown Plantn
Hartshole Fm
Broadwell
Double Waters
Walkham
Hatch Mill
Weir
Denham Br
Fishacre Wood
Newhouse
Woolacombe
Leigh
Newton
Well Fm
Hole Fm
BERE FERRERS
Bere Ferrers
RIVER TAVY
Weir Pt
New Barn
Tavy Br
Warleigh Point
Warren Point
Industrial Estate
Birch Wood
Tor
Maddacleave Wood
Mine (dis)
Harewood
Power Station
Chy
Chy
Gawton
Mount Tamar
Tuckermarsh
Rumleigh
Mount Tamar
Bere Alston
Leeches
Ley Fm
New Park Fm
Thorn Point
Neal Point
Landulph
Skinham Point
Skinham
Kingsmill Lake
Well
Landulph Cross
Colloggett
The Rock
Morwell
Morwell Wood
Morwellham (Museums)
Slimeford
Oakenhayes
Harewood
Weir
Calstock
Cotts
Buttspill
Asheric
Helstone
Lockridge Fm
Weir Quay
Clamoak
Holes Hole
Clifton
Hooe
Cargreen
Haye
Wayton
St Anns
Sch
Grove
West Kingsmill
Moditonham Quay
Carkeel
Gunnislake
St Ann's Chapel
Drakewalls
Albaston
Norris Green
Chapel
Ward Ho
Cotehele Quay
Cotehele Ho
Bohetherick
Braunder
Whitsam
Hewton
Newton
Morden
Cleave
Burraton
Burcombe
Well
Haye
Chapel Fm
Halton Quay
Pentillie Castle
Crosspark Wood
Landulph
Botusfleming
Ellbridge
Bicton
Stockadon
Doghole
Motel
Broadmoor Fm
CALSTOCK
Harrowbarrow
Metherell
Treragin
Trehill
Tumuli
Honicombe
Inn P
St Dominick
St Dominick
Smeaton
Brendon
Tremoan
Botusfleming
Mount Ararat
RIVER TAMAR
Halton Barton
Tumulus
Weir
Milton Combe

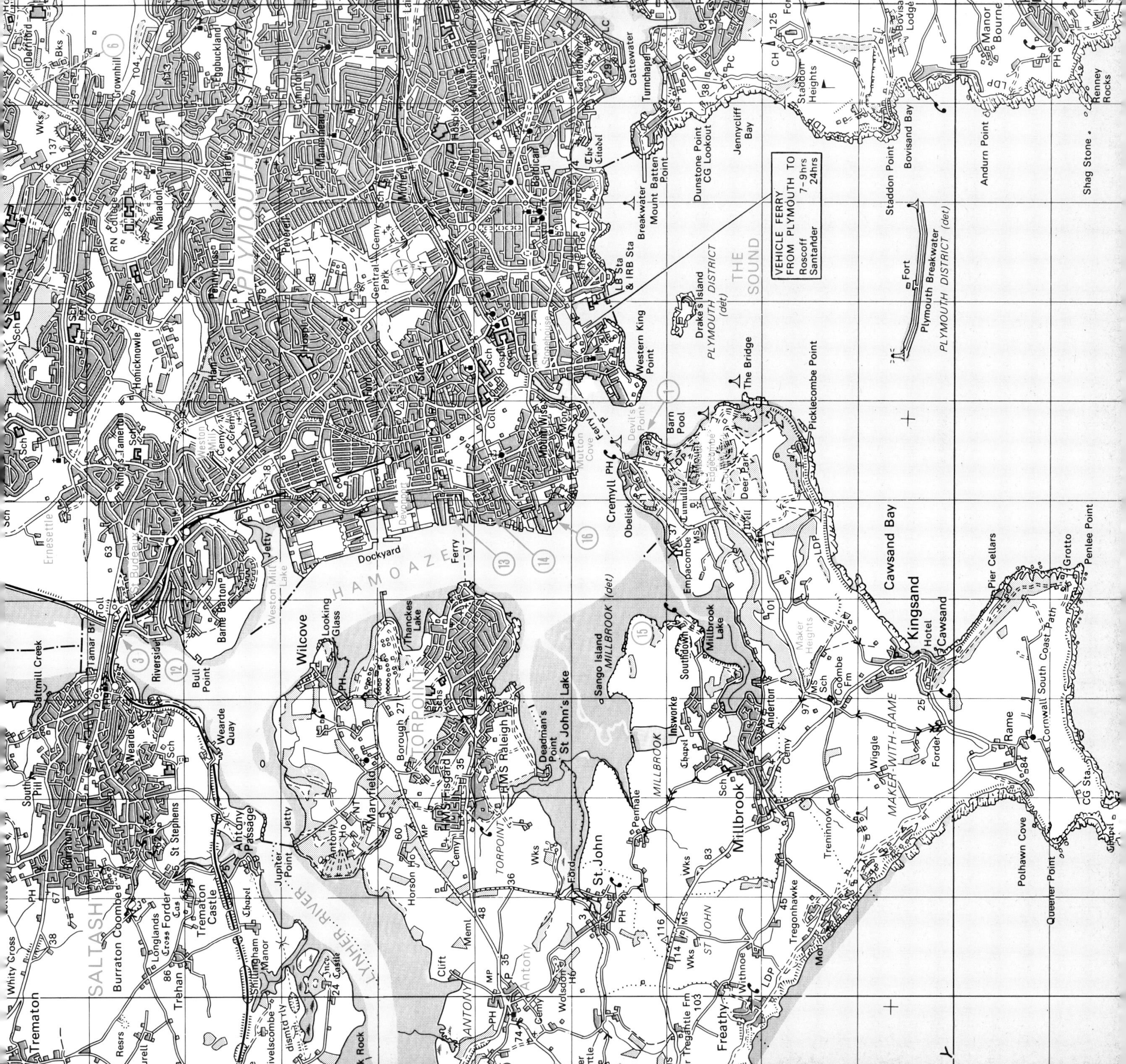

PLYMOUTH DISTRICT
SALTASH
TORPOINT
HAMOAZE
THE SOUND
LYNHER RIVER
TAMAR RIVER
CAWSAND BAY
MAKER-WITH-RAME
MILLBROOK
PLYMOUTH DISTRICT (det)
PLYMOUTH BREAKWATER
Plymouth Breakwater
Cornwall South Coast Path

VEHICLE FERRY
FROM PLYMOUTH TO
Roscoff   7-9hrs
Santander   24hrs

Trematon
Whity Gross
Burraton Coombe
Trehan
Longlands
Cross Forder
Trematon Castle
Antony Passage
Jupiter Point
St Stephens
Wearde
Wearde Quay
Bull Point
Riverside
Wilcove
Looking Glass
Thanckes Lake
Borough
HMS Fisgard
HMS Raleigh
Maryfield
Antony Ho
Horson Ho
St John's Lake
Deadman's Point
Sango Island
Southdown
Insworke
Millbrook
Millbrook Lake
Anderton
Penhale
St John
Wiggle
Treninnow
Forder
Combe Fm
Maker Heights
Kingsand
Cawsand
Rame
Polhawn Cove
Queener Point
Penlee Point
Pier Cellars
Grotto
Freathy
Withnoe
Tregonhawke
Tregantle Fm
Wolsdon Ho
Saltmill Creek
Ernesettle
St Budeaux
Weston Mill
Barne Barton
Weston Mill Lake
Dockyard
Devonport
Keyham
Ferry
Stonehouse
Stoke
Mount Wise
Mutton Cove
Cremyll
Obelisk
Devil's Point
Western King Point
Barn Pool
Mount Edgcumbe
Deer Park
Empacombe
Cremyll Ferry
Drake's Island
The Bridge
Picklecombe Point
Fort
Maker
The Hoe
The Citadel
Central
Cemy Park
Cathedral
Cattedown
Turnchapel
Mount Batten Point
Breakwater
Dunstone Point
CG Lookout
Jennycliff Bay
Staddon Heights
Staddon Point
Bovisand Bay
Andurn Point
Shag Stone
Renney Rocks
Bovisand Lodge
Manor Bourne
Manaton
Honicknowle
Manadon
Hartley
Compton
Lipson
Mannamead
Mutley
Beacon Park
Ford
Kings Tamerton
Saltash
Jupiter Point

Calstock Viaduct from the east

Tidal rivers

The artificial tidal limit of the Tamar is known as Weir Head, and the weir there may be as old as the fourteenth century. As far as the great bend at South Hooe, the estuary is riverine in character, gradually widening from thirty to three hundred yards wide. Its course is a fine example of superimposed drainage, whereby the river has cut down through rocks to form a steep-sided valley, in some places as much as five hundred feet deep, and almost gorge-like in places like the Morwell Rocks (4). This erosion must have occurred relatively fast on the geological time-scale for the Tamar has not had time to straighten out its bends. It flows in great incised meanders, the outer slopes covered in trees. The valley is thus one of great, and in places spectacular, beauty, compared by the Victorians with the gorge of the Rhine. The tributary valleys are likewise steep-sided. In almost all cases, the flood-plain on the inside of the meanders has been reclaimed by embanking. Outside the embankment one finds at first reeds, then at low water, a muddy slope extending down to the channel. The only bridge crossing the Tamar in this uppermost section is the railway viaduct at Calstock (5), an impressive structure built in 1906. Apart from the Tamar salmon fishery, the principal occupation in this area now is market gardening and farming. For many centuries, however, there was a thriving mining industry here.

The tidal limit on the Tavy is also artificial, but much more recent than the weir at Weir Head. In 1954 a dam was built at Lopwell, which cut off the estuary and created a freshwater lake over half a mile long out of the lowest part of the non-tidal Tavy. This lake is used as a reservoir for water supply; water is pumped from here to the treatment works at Crownhill (6). The riverine section of the Tavy estuary extends as far as Mount Jessop (7).

The middle reaches

The middle reaches include the stretch from South Hooe to Saltash Passage on the Tamar, and the remainder of the Tavy branch. They are much wider – five hundred to one thousand yards – than the higher sections. Their beauty is marred by the 400 kv electricity supply line, which crosses both branches on massive pylons, rather than in a tunnel. At high tide these reaches are wide expanses of water between saltings which are covered only at high spring tides. They are a haven of relative tranquillity, with only the occasional boat passing or train rattling over the bridges spanning the mouth of the Tavy (8) and Tamerton Lake (9). This railway was once the main line of the old London and South Western but since the Beeching axe it has been relegated to single-track and branch-line status; it provides the inhabitants of the Bere Alston peninsula with easy

The Tamar estuary at Weir Quay

access to Plymouth. In summer yachts lie to moorings off Weir Quay and Cargreen where there are sailing clubs. At low tide extensive mud-flats are uncovered, providing feeding grounds for many species of birds. Tamerton Lake is now the only major tributary on the Devon side, but formerly there were large creeks west of Horsham and south-west of the Ernesettle housing estate. The southern part of these middle reaches, below the Tamar – Tavy confluence, is dominated scenically by the two great bridges: the Royal Albert Railway Bridge, completed in 1859 by Brunel, and the Tamar Road Bridge, opened in 1961.

The Hamoaze

South of the Royal Albert Bridge the estuary deepens considerably and a trench varying in depth from thirty-six to one hundred and thirty feet runs the length of the Hamoaze and into Plymouth Sound. This may have been cut during the last Ice Age when sea-level was some three hundred feet lower than at present.

The Devon shore of the Hamoaze is part of the city of Plymouth and largely artificial. Formerly, three large creeks existed, but embanking and infilling have caused one to disappear completely and the other two to be severely reduced in area. These were Weston Mill Lake, which once ran up to the mill itself, Keyham Lake, which was where St Levan Road now is, and went as far as the eastern viaduct (**10**); and Stonehouse Lake, which once extended as far as Pennycomequick (**11**). (This name, incidentally, is English, and any attempts to explain it as Celtic are bogus.) Stonehouse Lake was reclaimed in two sections: first to Millbridge (now Victoria Park), and thence to Stonehouse Bridge, now a bridge no longer. One result of this extensive reclamation is that, on a small scale, the eastern bank of the middle Hamoaze appears as a smooth curve, and the channel roughly uniformly wide (about six hundred and fifty yards). Further to seaward, the estuary reduces to half this width, as it cuts its way through the ridge of Plymouth limestone in a tortuous channel which has always been difficult to navigate.

River-borne commerce

In an area with poor road communications, the river was for centuries the major highway. The Danes sailed up the Tamar in 997 and attacked Lydford; five centuries later, the Edgcumbe family used the river as the means of communication between their two houses at Mount Edgcumbe and Cotehele. The estuary was the principal outlet for the products of the two occupations, farming and mining, which are described below. The merchandise was at first carried by large open boats each with a single mast and sail, known as market boats. By about 1840, a distinctive type of sailing vessel known as the Tamar Barge had evolved, and these craft were the primary means of commercial transportation until the late 1930s, when they were replaced by powered vessels or road transport. Fortunately, an example has been preserved and restored by the National Trust: the *Shamrock* lies at Cotehele Quay, on the Cornish side. Only in the last fifty years has the Tamar lost its importance as a route. Above the confluence with the Tavy all waterborne commercial traffic has disappeared and the only vessels to be seen are pleasure craft.

The principal quay on the Tamar was Morwellham, which dates from about 1240 and which in the mid-nineteenth century was a busy port. There were at least twenty-five other quays on the Devon side, now disused and, in many cases, disappeared altogether. Those accessible by car are Hole's Hole, Weir Quay, Clamoak, Lopwell, Gnatham, Maristow, Ware's Quay (**12**), Pottery Quay (**13**) and North Corner (**14**).

Cotehele Quay, with the Shamrock

Agriculture and horticulture

In the Tamar Valley there is a rich historic diversity of farming mirroring the many ancient houses and settlements that are dotted along the valley from source to mouth.

There is no doubt that the steeper banks at the sides of the River Tamar and the River Lynher were once heavily wooded, as indeed they still are at Morwell and Cotehele with plateau and flatter areas being cleared for agriculture in the past.

Dairy farming together with cattle- and sheep-rearing predominate but there has been an increasing amount of corn and arable cropping in more recent years. Historically, it has always been an area of small farms and up to about twenty years ago the majority of them were less than a hundred acres. Recently, however, farms have amalgamated, owing in part to government policy and also to economic survival, so that the majority of farms in this area are now probably well over a hundred acres with several in-hand estate farms exceeding five hundred acres, as at Antony, St Germans and Pentillie.

With the increasing size of farms and the switch to arable farming the small intricate interweaving of hedgebanks dividing small fields is now fast disappearing; up to about 1975 it was not uncommon to find farms which had half their acreage in fields of less than three acres, but now with larger machinery and government subsidies these fields have been made into ten- to fifteen-acre blocks and the character of the valley is changing.

However, the Tamar Valley in its tidal parts is very fortunate in still having landed estates on both sides of the river – the Cotehele estate at Calstock, the Duchy of Cornwall at Landulph, the Pentillie Estate above Cargreen, the Ince estate along the northern fringes of the Lynher estuary, the St Germans estate at Port Eliot at the west end of the Lynher estuary, the Antony estate along the southern fringes of the Lynher estuary, the Roborough and the Warleigh estates, both around the Tavy estuary. Although most of these estates have land in hand in their home farms, the majority of the land is still let out on tenancies. The tenant farmers have to abide by the tenancy agreements and cannot take out hedges, cut trees, fill in ponds and materially alter their holdings without the estate owners' consent and it is quite evident that these constraints have slowed down the rate of change in the valley and in many cases the tenanted farms are the best kept in the area.

Moreover, all these estates have their principal houses which are often surrounded by parks and, in addition, many have woodlands, some for shooting purposes, which have been well maintained enabling the Tamar Valley to retain its essential character of long ago. The hanging woodlands between Gunnislake Bridge and Calstock are an interesting feature, as are the woodlands surrounding Cotehele House, with another delightful wood flanking the river on the Devon side opposite Cotehele and Halton Quay. Below Halton Quay are the incised valleys clothed in woodland lending privacy and interest to Pentillie Castle. Likewise, the lower reaches of the Tavy have deep wooded valleys running into the estuary. St Germans is more park-like, as is Antony, but there are many delightful small woods around Tamerton Lake, Moditonham Creek and St John's Lake opposite Devonport. Fortunately, most of these woods are deciduous and it is not until Morwell on the Devon side that there are large blocks of conifers. Even here the Earl of Bradford has been conscious of the effects of conifers on the landscape and considerable effort is made by the estate to lessen the impact of the normally uniform blocks of coniferous woodland.

Another particular feature of agriculture in the lower reaches of the Tamar Valley is market gardening. The favourable climate, with high levels of sunshine, moderate rainfall and good, deep, well-draining soils derived from the Devonian 'shillet', encouraged market gardening to flourish at the end of the last century. The growing population of the 'Three Towns' of Plymouth, Devonport and Stonehouse (amalgamated in 1914) provided a ready market for produce.

Hundreds of acres of woodland, mainly on steep-sloping ground, were cleared in the late 1800s and market gardening has flourished on these steep slopes ever since, with many of the same families being involved for several generations. In the Calstock and St Dominic area boats used to take the produce down to Plymouth and bring back dung which would be spread on the land to increase fertility. With the coming of the railways it was possible for St Dominic market gardeners to go down to Cotehele Quay, row across the river and take their produce up to Bere Alston station: in some five hours' time the produce could be on sale in London. Strawberries, anemones, flowers, *Pittisporum*, vegetables and soft fruit were grown in abundance and much of the valley was planted up with orchards of cherries (including local varieties of Burcombes, Fices and Birchenhayes), apple and cider trees, together with plums. Many of these orchards are now gone, the trees ripped out and winter wheat or barley growing in their place. Few orchards are maintained and indeed most of the old varieties of fruit trees are almost extinct. But in spring time the few old remaining orchards come into their own with marvellous displays of blossom.

Most of these market-garden holdings are on steep ground in the little valleys that run down to the river; many of them are hidden from view, well sheltered from the prevailing westerly winds. Times are changing and horticulture is no exception. These steep market-garden holdings are demanding both in labour and in machinery and as demand slackens for fresh produce so there are fewer younger people coming forward to take on these smallholdings. Up and down the valley it is a common sight to see the outline of former market-gardening strips now left idle and untended. On the Cotehele and Pentillie estates many of these former steep holdings which once flourished and were worked from dawn to dusk are now returning to woodland, so in one hundred years full circle has been reached.

Warleigh House beside the Tavy estuary

The valley is no different from most other parts of the country with regard to farm buildings. Many an uninspiring prefabricated cattle yard or general purpose building has disfigured the landscape through being badly sited, poorly designed and of material that is out of keeping with the area. There are signs of improvement now with the use of timber cladding, lower-profile roofs and an attempt to set the buildings in the landscape rather than upon it. Again, the farm buildings on the landed estates are generally more attractive and blend in better with the older traditional stone and slate farm buildings. Unfortunately, these latter buildings are largely unsuitable for modern farming practices and it is not always easy to make satisfactory use of them. Recently, however, many have been converted into domestic dwellings and light industrial or craft use. As planning authorities become more conscious of landscape and historical-building values, many of these conversions are now carried out to a high standard and have an attractive finish.

Buildings on the estuary

The Tamar, of course, is the boundary between Devon and Cornwall and it is very difficult to isolate the buildings on the Devon side and view them only in a Devon context. Buildings on each side give value to those on the other in the whole estuary scene.

The Tamar at its lower end is dominated on the Devon side by Plymouth and the Royal Dockyard. The estuary has been put to warlike purposes since early times. Already, as early as 1287, a fleet was assembled at the mouth of the Tamar for an expedition against Guienne. Although it is now equipped to deal with a fast-moving modern war, such as the Falklands campaign, the remains of the seventeenth-century dockyard can still be seen from the river, together with many more buildings from the eighteenth and early nineteenth centuries, added as the navy grew and the dockyard became larger.

Torpoint, on the Cornish side of the lower estuary, expanded as the dockyard expanded during the Napoleonic wars and the names of little streets of late-Georgian terraced houses, such as Salamanca and Wellington, reveal what was preoccupying people's minds when these terraces were built.

At Torpoint, at the water's edge, can be seen a fine buttressed warehouse of the eighteenth century. This is now well preserved and in modern use. It has been the subject of several planning enquiries.

On the Cornish side of the mouth of the estuary, providing a green backdrop, are Maker Heights, surmounted by the handsome tower of the late-medieval Maker Church. Lower down the same slope is Mount Edgcumbe House. This home of the Earls of Mount Edgcumbe was burnt out by an incendiary bomb during the Second World War and rebuilt within the sixteenth-century shell after the war.

In 1966 the Central Electricity Generating Board proposed to build a power station at Insworke Point (15), a prominent tongue of land on the Cornish side between Maker Heights and Torpoint. This proposal was bitterly contested for some eight years. The plan is now in abeyance. If the plan had materialized, or ever did, it would profoundly alter all the characteristics of the Tamar estuary.

The ancient town of Saltash on the Cornish side, opposite St Budeaux in Devon, is much altered. It was badly bombed during the last war. The little medieval house where Sir Francis Drake's first wife, Mary Newman, lived still survives here in Culver Road. It has been restored and is open to the public. Across the river, high up on each side, the fine churches of St Budeaux and Saltash face each other.

The Tamar, like all great rivers, has been a highway through the ages. On its banks and the banks of its tributaries people came and settled and left their traces. The Iron Age people came and left their ring forts. The Danes came and left very little except their names. The Saxons, in turn, regarded the estuary as a political and military boundary, but did not cluster on its banks, on the Devon side, as the Cornish did on their side, except later at Bere Ferrers. All the way up, the little settlements, now expanded, survive on the Cornish side interspersed with the houses and estates of the great gentry, notably Cotehele and Pentillie.

Perhaps on both sides of the upper tidal river the Tamar is best remembered for its deserted tin, copper and arsenic mines. The old engine houses and their chimneys hang over the river in the most unlikely positions. In their time, these mines brought a degree of prosperity to the small towns, particularly Gunnislake and Calstock, where some houses of modest sophistication were built in the late-Georgian period.

Possibly the outstanding features of the river are the two or three great medieval bridges that cross it in its upper reaches.

The mining industry

The Tamar flows through the easternmost part of the great Cornish metalliferous belt, which extends just into Devon. In this belt the central lodes run east–west and are mainly of tin and copper, for example at Devon Great Consols north of Gunnislake; the peripheral lodes run north–south and are mainly of lead, silver and arsenic, as at North Hooe. Mining began on a small scale in the twelfth century, the two outlets being down the estuary or overland to Tavistock, an ancient stannary town. It developed greatly after the Industrial Revolution and reached its heyday in the mid-nineteenth century. Extensive ancillary and other industrial undertakings were established: a smelting works near Weir Quay, quarries and shipbuilding yards, an arsenic refinery at Rumleigh, brickworks, ropeworks, and limekilns on many of the quays. Air pollution was common.

The collapse of the mining industry was remarkably swift. In a generation, from 1870 to 1900, external economic factors caused most of the mines to close.

Today, traces of the industrial past lie everywhere around the upper Tamar estuary. In the 1960s Morwellham was a strange place, ruined and overgrown. The current interest in industrial archaeology had not then arisen and this abandoned port lay 'undiscovered'. Nowadays it has a thriving tourist centre, administered by the Dartington Amenity Research Trust, which is a must for anyone interested in this aspect of the Tamar's history. The great dock, built in 1859 to take up to six vessels, each of three hundred tons, has been restored. Visitors can study old photographs and even take a ride on a train in a reopened mine tunnel.

The dockyard at Devonport

The recent history of the Hamoaze is that of the naval dockyard at Devonport (formerly Dock). This was established at Point Forward **(16)** in the 1690s, during the reign of William III, whose statue stands just to the west of Mutton Cove. During the years 1718–25 the first major extension was carried out, taking in twenty acres of land to the north of North Corner. Known at first as the Gun Wharf, it was renamed Morice Yard in 1939. The original yard was enlarged in stages from five acres to seventy acres, and today this forms the historic part of South Yard. The change from sail to steam brought about the construction of a further yard of sixty acres to the north of Pottery Quay which was finished in 1853 and called Keyham Steam Yard. Thus each major expansion moved further north. The greatest of all was that of 1896–1907, one hundred and fourteen acres in all, which brought the northern limit to the mouth of Weston Mill Lake. During the 1970s there were three major developments: the Fleet Maintenance Base, the Submarine Refit Complex and the Frigate Complex, all in North Yard. This last is the most obvious since its huge building, capable of housing three frigates at a time, can be seen for miles around. The total area of the dockyard is now nearly three hundred acres.

Wildlife

Because of its unique situation the estuary is a haven for wildlife and attracts many migrant birds. The mud-flats are the largest in area in the south-west. In the winter waders can be seen at low tide feeding at the water's edge, wigeon and Brent geese feed on the eelgrass growing in the sea below the tide level, dunlin can be seen roosting in flocks at high tide and a cruise along the Tamar will provide the observer with a fine view of the avocets feeding on small shrimps. The salt-marshes, streams and river banks also attract the curlew, lapwing, oystercatcher, greenshank and redshank, plover, knot, turnstone, godwit, heron and gull. Sometimes unexpected visitors, such as the spoonbill, semi-palmated sandpiper, egret and osprey, are to be observed.

Wildfowl, such as shelduck and mallard, nest in the banks, kingfishers feed along the streams and swans muster in large numbers. Sheviock Wood supports many breeding birds, including heron, buzzard and tawny owl. Many animals live in the mud – sandhoppers, lugworms, and crabs – which attract the birds. Rabbits, squirrels, frogs and toads thrive in their quiet habitat. Salmon and sea trout pass through the estuary to their spawning ground.

The tidal estuary supports many plants which can survive the highly salty conditions. These include reeds and sedges, glasswort, sea purslane, sea aster and sea lavender. Grasses include cordgrass, sea arrowgrass and eelgrass. The river banks abound in many wild flowers and a wide variety of trees and shrubs. A catastrophe occurred when Dutch elm disease destroyed so many elms recently.

The estuary and its environs provide a safe environment for all kinds of wildlife, which it is not only our pleasure to enjoy but our duty to preserve.

Scientific investigation

Before about 1970 there had been relatively little scientific investigation into the estuary but increasing interest in environmental pollution caused the Marine Biological Association at Plymouth to begin an intensive research programme on the estuary. This was partly brought about by the threat, now mercifully receded, to construct a power station on the banks of the estuary. In recent years the Association has been joined by two other institutions, Plymouth Polytechnic and the Institute for Marine Environmental Research, so that the Tamar is now under almost constant scrutiny. Scarcely a day goes by when a survey boat from one or other of these institutions is not seen somewhere in the estuary, collecting samples for analysis.

Both individually and in co-operation, they have investigated aspects ranging from the fauna of intertidal muds to the concentration of trace metals. The latter are of particular interest because of the mineral deposits from the metalliferous belt. Detailed hydrographic surveys have been made, revealing deep pits in the bed of the estuary. Numerical models of varying complexity have been used to simulate the tides, the circulation and the movement of sediment.

King William IV Victualling Yard, Plymouth

Access

Public access in the Tamar Valley is limited and in particular access to the river itself. This limitation of access is the saving grace of the river from source to mouth and especially in its most beautiful reaches from Greystones Bridge past Horsebridge and in the tidal reaches below Gunnislake Weir. It is only in these aforementioned places that the river is crossed apart from the latter-day Saltash Bridge and there are very few tracks or roads that run up and down the valley. Little byways have to be sought out to find the small settlements bordering the Tamar. These are not the roads or the place for those in a hurry, but to those with time and genuine interest there is much to be admired and learnt up and down the valley. Settlements such as Millbrook, St John, Antony Passage, Cargreen, Bere Ferrers, Weir Quay, Halton Quay, Cotehele Quay, Calstock, Tucker Marsh can all tell their story but no single road connects them.

Undoubtedly the best way to see the valley in its true splendour is to take a steamer from Phoenix Wharf on Plymouth Barbican and then to admire the otherwise hidden delights of the river up to Calstock and beyond. For those who would prefer something different on the way back, a train can be taken from Calstock, through the Bere Alston peninsula, over the River Tavy at Warleigh and back to Plymouth.

Even footpaths and bridleways are scarce but the 1:50 000 scale Ordnance Survey map is still a passport for the discerning walker to discover the delights of this wonderful valley.

This chapter was prepared by Dr K. J. George, Mr O. Prattent, Mr Jack Spiers and Miss Chris Goodall on behalf of the Tamar Protection Society. The Society was formed in 1967 to encourage and support the enlightened use of the natural resources of the area, safeguarding its unique character, natural beauty and amenity value, and to ensure that any development is of a sympathetic nature and in the best interests of the area itself and the country as a whole.

After many years of painstaking restoration work by the Society's members, the cottage of Mary Newman (first wife of Sir Francis Drake) has recently been opened to the public and has become a tourist attraction.

The Secretary of the Society is Chris Goodall, Herons Ford, St John, Cornwall PL11 3AR, to whom all enquiries should be addressed.

11
Taw ~ Torridge Estuary

The Taw and Torridge estuary is the only large estuary in the whole of the one-hundred-mile north Devon coastline, and, as such, historically it has been the hub of communications for the area and has played an important role in Britain's seafaring tradition. Before the coming of the railway in the middle of the nineteenth century the sea was the main highway and generations of seafarers have sailed in and out of this estuary, many in very small ships, to bring life and prosperity to this corner of our island.

Every town, village, and hamlet along the coast and the estuary had its own limekiln, with the lime and coal being shipped from Wales to make fertilizers for the land. As you look around the shores of the estuary you will see the remains of these kilns every mile or so along the shore.

Life has changed, however, and the estuary is no longer the highway that it once was. In 1851 people going to London to see the Great Exhibition at the Crystal Palace went first by sea to Bristol

The Taw estuary, with Barnstaple in the distance

Velator Quay on Braunton Pill

and then by one of the 'new' trains to London. Now the railways have all but disappeared from this region, shipping since the Second World War has declined dramatically and, though Bideford is still a port for small coasters, the activity is slight and the economy relies on the 'juggernaut' lorry.

With this background in mind let us take a journey round the estuary, starting in the north with Braunton and finishing south of the Torridge at Westward Ho! For those who wish to make an actual journey on foot, there is a footpath from Braunton to Barnstaple along the River Taw; this will shortly be extended along the other side of the river to Bideford, following the route of the disused railway.

Braunton was at one time 'Brannocminster', named after St Brannoc who brought the Christian message from Wales and formed a settlement here in A.D. 300. It has always been a large centre of population. The River Caen, perhaps named by some homesick Norman, runs into the Taw through Braunton Pill (1). Every winter until the Second World War the Pill was a forest of masts as the many sailing ships which plied the coast came in for repairs. Braunton had sail lofts and all the necessary skills for maintaining these ships. If you go down the Pill and along the Toll Road (2) through Braunton Marshes you will come to what is now called the White House but was previously the Ferry House for Appledore and Instow. Turning westwards from the White House brings you to the

Saunton
Lobb
Buckland Ho
Marwood
Wigley Cross
CH
Fairlynch
St Brannock
Chapel
Whitehall
Lee Ho
Guineaford
Kennacott
Somerset and North Devon Coast Path
Ash Barton
Pippacott
Prixford
Kingsheanton
Golf Links
Braunton
Tower
Braunton Down
Luscott Barton
Waterlake
Mainstone
Varley
Saunton Sands
Sandy Lane Fm
Park Fm
Windy Cross
Horridge
West Ashford
Springfield Cross
Blakewell
Tutshill
Braunton Great Field
Velator
Heanton Punchardon
Heanton Ho
Heanton Court
Ashford
WEST PILTON
Westaway
Braunton Burrows
Marstage Fm
Wrafton
TOLL
Chivenor Airfield
Strand Ho
Pilland
Bradiford Ho
Range
Braunton Marsh
Chivenor
Penhill Point
Course of old railway
Pilton
Bradiford
RIVER YEO
Dunes
Horsey Island
Allen's Rock
Quay
BARNSTAPLE
Pottington
Weir
Airy Point
White Ho
Saltpill Duck Pond
Penhill
Hollowcombe
Sticklepath
Nature Reserve
Stone Row
Muddlebridge
The Clampitts
Range
RIVER TAW
Power Station
Lower Yelland
Inn
Frenington
Combrew
Bickington
Herton
The Neck
Crow Rock
Crow Point
Cemy
Yelland
Horsacott
Bishop Tawton
Instow Sands
Bickleton
Myrtle Cottage
Brynsworthy
Roundswell
Appledore
Instow
Worlington
Knightacott
Lydacott
Collacott
Rookabear
Abattor
Upcott
Hollamoor Clump
Tower
Little Pill
Hotel
The Quay
Raddy
Orchard
Netherby
Hollamore Fm
Tawstock
Knapp Ho
Bloody Corner
Tapeley
Combe
WT Sta
Brookham
Litchardon
Nottiston
Eastacombe
Holy Well
Park Gate
NORTHAM
Huish
Huish Moor
Knowle Fm
Holmacott
St John's Chapel
Stonyland
Hillside
Corffe
Deerpark Lodge
Burrough
Inn
Trayhill
Pyewell
Voscombe
Collabear
Uppacott
Smemington
Westleigh
Eastleigh Manor
Lovacott Green
Prospect Place
Charlacott
TAWSTOCK
Fire Beacon Cross
Orchard Hill
Ball Hill
Bradavin
Eastleigh
Horwood
East Barton
West Barton
Tennacott
Linscott
Yellan
Rollestor
Southcott
Newton Cross
Sideham

CG
Lookout
Crow Rock
Crow Point
RIVER
19
20
MS
T
Cemy
Yelland
Horsacott
51
79
78
Bickleton
Myrtle Cottage
83
Lydacott
Rookabe
Ab
Instow
The Barton
Worlington
Knightacott
Collacott
107
100
Golf Links
Sandymere
PC
Northam Burrows
Skern
LB Sta
Appledore
Instow Sands
Tullingcott
Orchard
74
Brookham
WT Sta
Mus
P
Ferry
Hotel
The Quay
Raddy
Litchardon
103
38
Pitt
P
P
5
P
Huish
98
113
122
Huish Moor
79
Goosey Pool
Diddywell
58
3
124
WESTWARD HO!
CH
Knapp Ho
Shipyard
MS
Combe
119
Holmacott
120
Voscombe
Rushco
CG Sta
Bloody Corner
Tapeley
Knowle Fm
Pyewell
27
NORTHAM
Trayhill
Lovacott Green
ck Nose
Coast Path
Burrough
Inn
P
Westleigh
115
id's Pool
83
North Devon Coast Path
Buckleigh
5
Eastleigh Manor
74
Horwood
25
Pusehill
MS
Ball Hill
Bradavin
Eastleigh
East Barton
Cornborough
Silford
20
9
Southcott
103
West Barton
Course of old railway
Orchard Hill
Park Fm
125
Kenwith
Castle Mound
5
Ashridge
Lower Lovacott
am
Rickard's Down
23
Combe
Weach Barton
Coll
Cemy
Pillhead
MS
55
Combe Walter
BIDEFORD
Hosp
Salterns
East-the-Water
Little Pillhead
Webbery
Buddacombe
Bartridge Common
81
Rixlade
Inn
Hospl
Abbotsham
P
Winsford
Moreton Ho
M
Bulworthy
Handy Cross
Hos
Warmington
Woodtown
89
121
Ford Ho
Gammaton
T
Bowood
102
MP
75
Stony Cross
Adjavin
Caddsdown
105
Upcott
Gammaton Resrs
145
Alscott Barton
Alverdiscott
Ba
High Park
150
Windmill Cross
Tennacott Fm
Beara
ON
Jennett's Resr
Hallsannery
Gammaton Moor
ALVERDISCOTT
Winscott Barton
117
Littleham Court
101
7
Pillmouth
Oldiscleave
Haddacott Cross
South Moor
Ashridge
97
LANDCROSS
Brownscombe
Garnacott
Moorhead
102
Landcross
Huxhill
Guscott
Huntshaw Water
Hill Fm
Yeo Vale
P
Heale Ho
T
Hallspill
Little Weare Barton
Netherdowns
Twitchen
Sout
Rollstone
Hooper's Water Fm
T
HUNTSHAW
River Yeo
105
Edge Mill
West Annery
Venton
Huntshaw
161
Bulland
Annery Kiln
Knockworthy
th Yeo
The Hill
22
Park
Southcott
Berry Castle Fort
Foxes' Cross
Waggadon
Orleigh Court
Hall
TV Sta
Beara
Orleigh Mills
Saltren
Weare Giffard
Footlands
Delve's Grove Cross
Haycroft
Halsbury
80
Looseham
Orchard P
Higher Darracott
Darracott Moor
Tumuli
Bowden
Upcott
Ley
Downes
20
141
Priestacott Moor

Aerial view of Braunton Great Field

Nature Conservancy Council's car park **(3)** with its information boards of the area known as Braunton Burrows. This is a nature reserve and also an army training area, though now only occasionally used for the latter purpose. The car park can also be reached from Braunton by the 'American' road **(4)** so-called because it was made by American forces during the Second World War. The beaches were then covered with defences to form a replica of the German defences on the beaches of Normandy and used for training countless Allied soldiers for the landings in June 1944. (Warning: the road has not been maintained and it is better to leave your car at Sandy Lane car park **(5)**.)

The Burrows have some of the largest natural sand dunes in Europe and the area is a nature reserve of international repute. In June and July it is a carpet of flowers. Some plants such as the tiny French toadflax grow nowhere else in the British Isles, while others which are rare elsewhere grow here in abundance. Among these are the water germander, the sea stock and the round-headed club-rush. You do not have to look far to find the attractive little yellow seaside pansy, while the round-leaved wintergreen, first found here in 1964, has spread rapidly over the whole area. Several species of orchid can be found, pyramidal and southern marsh are both common, while marsh helleborine is abundant in every wet slack. Less easy to find is the little brick-red sub-species of the early marsh orchid, sometimes called the dune orchid. The Nature Conservancy Council has produced a Flowering Plant List which can be obtained from the Warden.

A wooden walkway takes the visitor over the dunes to the beach, coming out near what was until recently the site of a lighthouse **(6)**. At low water numerous rock pools and gullies are exposed which are the homes of a variety of small fish, prawns, crabs and anenomes. If you catch any of these little creatures just for fun or to show the children, then please put them back in their own pools, since in a strange environment they may die.

Again, at low water you can scratch for cockles in the sand or during the season watch the salmon fishermen laying out their nets, rowing out in a long circle and hoping for a good catch. You may be upset, however, to see the salmon stunned by blows on the head or the rough treatment meted out to crabs whose legs get entangled in the nets.

Going northward along the beach brings you to Airy Point where cormorants love to congregate and digest, facing the wind with their wings spread. On round the corner you come to Saunton Sands, with two miles of beach and surf, once the scene of land-speed-record attempts before motor cars became too fast, and now a venue for surfers, windsurfers and landyachts. If you look out to sea you can spot the buoys which mark the channel for ships coming in over the bar.

Braunton Great Field is situated between Braunton and Braunton Marshes and occupies some three to four hundred acres. The Great Field is a relic of Anglo-Saxon communal farming, the strips of land being handed down from generation to generation.

The old railway line from Braunton to Barnstaple is now a footpath and cycle track. Once past the Royal Air Force training base at Chivenor, the path takes you along the north bank of the River Taw, a shallow river with sandbanks and large areas of salt-marsh. Typical vegetation is sea couch grass, sea wormwood, sea spurry and sea meadow grass which, with many other species, provide natural over-wintering, breeding, feeding and resting areas for a great variety of birds. Most of the estuary is classified as a Site of Special Scientific Interest and the whole is of international importance, being second only to the Exe estuary in the south-west of England. Some species to look for are curlew, ringed plover, redshank, dunlin, oystercatcher, teal, lapwing, golden plover, shelduck and godwit.

Continuing towards Barnstaple, on the hillside is Heanton Punchardon, named after William de Punchardon who was granted the land by King Richard I. The churchyard contains the graves of many airmen who lost their lives flying from Chivenor during the Second World War and of others since.

Heanton Court, on the riverside, is now a restaurant but was once the home of the Bassett family, one of Devon's oldest families, and at one time had the reputation of being haunted. Another fine house is Upcott, on the hillside on the Barnstaple side of Ashford. Built in 1710 by the Harding family of Combe Martin, it is now divided into flats but is still a fine example of the small English country house of the period.

Ashford is a pretty village which has now become a dormitory area for Barnstaple. If you go inland to Marwood, some two miles distant, you will find, close to the lovely old church, a remarkable garden which is open to the public. In spring the camellias are a delight, but at any time there is much of interest, including a great variety of trees, a water garden and a collection of trees and plants from 'down under'.

Pilton, though now a part of Barnstaple, has a history, identity and character of its own. Its most attractive High Street leads up to the alms house and interesting church and the whole takes the visitor back in time to the early years of this century and away from the land of superstores. Pilton House is built on the site of the old priory founded by King Athelstan early in the ninth century.

On the other side of the River Yeo is Barnstaple. It is claimed to be one of the oldest boroughs in England and, with Pilton, was of importance in the early tenth century, being ordained by Athelstan to be one of the four strongholds in the west to defend the region against attack from the sea. The Castle Mound **(7)** dates from Norman times and is situated opposite the Civic Centre which certainly does not! The Castle was probably built by Judhel of Totnes, again as a defence against attack by sea, but only the mound remains. Barnstaple rose to commercial importance as a wool town during the reign of Queen Elizabeth I, trading with France and Spain. It equipped two ships against the Spanish Armada.

There are many places of interest in the town: the parish church with its crooked spire **(8)** and the old houses nearby; the Long Bridge **(9)** first built some six hundred years ago; the alms houses

Barnstaple Bridge

(10) which have bullet marks from the Civil War when Barnstaple sided with Parliament; the big covered market on Fridays and Butchers Row (11) opposite. Poet and playwright John Gay was born here. At one time Barnstaple had its own assay office and Barnstaple silver is greatly prized; examples of it can be seen in the Town Hall.

Evidence of Barnstaple's seafaring life may be found in the Customs House on the Strand, which has a fine statue of Queen Anne (12). Until the middle of the nineteenth century sailing ships were being built in the shipyards, the most famous being Westacott's. Many emigrants sailed in these ships to the New World and Australia, up to one hundred and thirty people being carried in each ship of only four hundred tons. Because of the shallowness of the river only small ships of up to fifty tons customarily traded to Barnstaple.

The railway reached Barnstaple, from Exeter, in 1856, built by the North Devon Railway and Dock Company. The line was broad gauge, but when it was taken over by the London and South Western Railway Company in 1861 the track was modified by the addition of an extra rail to take standard-gauge engines. For the next thirty years goods trains ran on the broad gauge while passenger trains used the standard gauge. The broad gauge was abandoned only in 1892. In the heyday of the railway there were four stations in Barnstaple. Now only the 'Junction' remains and is the end of the line (13).

Further up the river is Bishop's Tawton. Above the village is Codden Hill (630 ft), one of the ancient Beacon Hills. It is a good walk to the top with rewarding views of the surrounding countryside as far as Exmoor, Dartmoor and out to sea (the next beacon was Dunkery on Exmoor).

On the west side of the river, opposite Bishop's Tawton, is Tawstock. Set in lovely woodlands, Tawstock Church (14) is worth a visit and is a pleasant walk by footpath from the river. St Michael's School now occupies the fine eighteenth-century castellated country house, Tawstock Court, since medieval times the seat of the Earls of Bath and the Wrey family. Not far above Tawstock is New Bridge, the tidal limit of the river.

While it is possible at present for the venturesome to walk all the way from Barnstaple to Bideford, following the river, the disused railway line is being acquired by the County Council and made into a footpath and cycle track. This will make it possible for the less intrepid to enjoy the openness of the river and all its wildlife.

Fremington lies some three miles from Barnstaple. From Muddlebridge (on the main road), once the site of a pottery, it is a pleasant walk down Fremington Pill (15) to Fremington Quay (16). In the heyday of the railway this was a busy port with ships loading and unloading from waiting goods trains. Now it is the site of an abattoir and not so pleasant.

However, across the iron railway bridge there is the remains of a very large limekiln and you can walk along the shore or back the other side of the Pill to Fremington village. During and after the Second World War, the large red-brick early Victorian mansion (17) was the School for Combined Operations. Though the mansion is no longer Ministry of Defence property, there is between it and the river an army camp used primarily for adventure training of young soldiers. Between Fremington and Bickington and to the south of the latter village are the extensive clay pits (18) which over the centuries have supplied Barnstaple's famous potteries. In the early days of the American colonies much local pottery was shipped there.

Continuing westwards we come to Instow, a corruption of Johnstowe from the dedication of its twelfth-century church to John the Baptist. Until the latter half of the nineteenth century the village was situated on the hill behind the church, with just a few fishermen's cottages by Instow Quay, built in 1630. With the advent of the railway an effort was made to turn Instow into a spa and the large houses along the front, known as Bath Terrace, were built, along with a Bath House. Happily this development failed to expand, but even so Instow is now a village by the river. At the confluence of the Taw and the Torridge it holds many attractions. One of the earliest of well-known cricket clubs, the North Devon Cricket Club, founded in 1823, is sited at the end of the broad sandy beach and sand dunes (19). It has an attractive thatched pavilion. Nearby, the Royal Marines have a trials and training unit which accounts for the various landing craft and vehicles appearing on the beach, to the immense enjoyment of small boys.

A centre for sailing and various water sports, the North Devon Yacht Club is now situated on the site of the disused railway station.

Tawstock Church

Bideford Bridge

During the summer a ferry operates between Instow and Appledore for about three hours before and after high water.

Westleigh is a charming and still unspoilt village overlooking the River Torridge. The church houses some interesting old pews and memorials. Nearby is Tapeley House with its extensive gardens and interesting interior; it is open to the public from Easter to October. In the spring the long drive from the Palladian-style lodge at its entrance is a feast of daffodils. At the top of the drive the trees give way to a splendid view of the estuary and out to sea. In the grounds in front of the house lies the fallen obelisk erected by John Cleveland in memory of his son who was killed at Inkerman, having survived the Charge of the Light Brigade. The estate is at present owned by the Christie family, better known to opera lovers as the owners of Glyndebourne.

Bideford and East-the-Water are joined by the famous bridge of twenty-four irregular arches. The bridge was originally built in the early part of the fourteenth century by public subscription in Devon and Cornwall under the auspices of Grandisson, Bishop of Exeter. The Bishop, being influenced by a miracle (the parish priest claimed that a rock had suddenly appeared in the river indicating where solid foundations could be laid), granted indulgences to all who contributed to the work.

From its earliest days Bideford was an important port and many were the ships built at East-the-Water where Bideford Black was mined and used all over the country for treating ships' timbers. Mines Road above the main road was the site of the workings. In the sixteenth and seventeenth centuries trade with Newfoundland and the New World assumed great importance, reaching its peak in the eighteenth century. As long ago as the sixteenth century Bideford had twenty-eight ships working the Newfoundland fisheries. Now the port is used mainly by small coasters bringing in fertilizer and exporting china clay.

Many famous seafarers have come out of North Devon but probably none is better known than Sir Richard Grenville, explorer and cousin of Raleigh. He lived in Queen Elizabeth I's time and from his ship *Revenge* fought off fifty-three Spanish ships, destroying four of them. Five ships, three of them Grenville's, sailed from Bideford to join Sir Francis Drake at Plymouth to repel the Armada. Guns from Spanish ships of that period can be seen in Victoria Park **(20)**.

The author, Charles Kingsley, whose statue stands on the river bank near Victoria Park, wrote much of his work in the reference library of Colonial House, now the Royal Hotel. The room has been preserved and contains rococo plasterwork and ceiling, and panelled walls. At the top end of Bideford, above Grenville Street, is a fine covered market **(21)**, held on Tuesdays and Saturdays.

Up-river from Bideford bridge the appearance of the Torridge changes as it flows through a wooded valley with steeply sloping sides. As in the Taw estuary there is plenty of wildlife; look out particularly for herons and kingfishers.

The tidal part of the river ends at Weare Giffard, a delightful, though straggling, village nestling on the hillside above the river. This is 'Tarka country', home of Henry Williamson's *Tarka the Otter*. Outstanding in this area is the Elizabethan manor house, Weare Giffard Hall **(22)**, in medieval times the home of the Fortescue family who now live at Filleigh near South Molton.

Returning towards Bideford along the west side of the river, the small hamlet of Landcross is passed. Here, the River Yeo (not to be confused with the one at Barnstaple) flows from its thickly wooded valley into the Torridge. On entering Bideford you will pass Ford House, the site of the ancient ford across the river. From Bideford it is possible to walk along the river to Appledore. However, if you go by road you will come first to Northam, a small town which runs into Westward Ho!, a latter-day invention of which more later.

In King Alfred's time the coast was frequently raided by the Danes. On one occasion Hubba, with thirty-three longships fresh from pillaging South Wales, landed in North Devon. It has been suggested that the site was Skern Bay, below Northam, and that he laid siege to Kenwith Castle (23). Odun, Earl of Devon, marched on Hubba and defeated him, possibly at a place now known as Bloody Corner. The encounter involved great slaughter. Leaving Northam on the road to Appledore you will see the stone memorial set up in the last century to mark the victory. Hubba was killed and buried on the shore under a very large flat stone from which comes the name Hubbastone.

Appledore – a port from the earliest times – has played a key part in the development of the area. Today its narrow and picturesque streets contrast with the huge covered building dock of Appledore Shipbuilders, one of the most modern and successful yards in Europe, building a variety of ships of up to eight thousand tons. To gain a real insight into the estuary's maritime history a visit to the Maritime Museum in Odun House (24) is a must. This small but fascinating museum sets out to show with models, photographs and items of the past how life has been lived here throughout the ages. Trawlers and fishing boats still unload at the quay and it is possible to arrange for fishing trips either by the day or tide to try to catch a delicious bass or perhaps just a mackerel. From the quay you should walk down the picturesque Irsha Street to West Appledore and the lifeboat station where visitors are always welcome. Dominating Appledore is the Holt built by William Yeo who, on returning from Prince Edward Island, was the driving force in developing Appledore's shipbuilding capacity in the mid-nineteenth century. The house is now divided into flats.

Beyond Appledore and to the west of Northam is Westward Ho! In 1863 a company was formed to develop the agricultural land between Northam and the sea. Charles Kingsley had recently published his novel *Westward Ho!* and it was decided to use this name, complete with exclamation mark, for the development. The aim was to build a fashionable watering-place. Hotels and villas sprang up, a golf course, swimming pool and pier all followed (though the pier was soon a victim to the sea). The United Services College was opened in 1874. One well-known ex-pupil, Rudyard Kipling, based his book *Stalkey and Company* on the school, which was housed in what is now Kipling Terrace. In 1901 the Bideford, Westward Ho! and Appledore Railway was opened, though the link to Appledore was not completed until 1908.

The First World War changed the face of the whole development. The railway was requisitioned in 1917 and engines, rails and rolling-stock shipped to France. Following the war, the grandiose hotels and villas went into decline and the resort developed into the popular and rather more gaudy place that it is today. The beach, however, is magnificent, with its great stretch of sand and the roar of the pebble ridge at high water as the surf breaks against it. North Devon Golf Club is close to the beach and Northam Burrows is now a country park where a useful information centre has an interesting display of local shells. If you follow the first part of the coastal footpath westwards along the old railway track to Cornborough Cliffs (25) and onwards you will find splendid coastal views.

There is much to be enjoyed in the whole of this region. Even where there are no footpaths, the Devon lanes, away from the main roads, provide excellent walks and the hedgerows provide endless variety and interest.

This chapter was prepared by Capt. C. J. A. Johnson on behalf of the Taw and Torridge Estuary Forum. Mrs Margaret Tulloh's advice on local flora is gratefully acknowledged. The Forum was formed in 1980 and consists of one councillor from each town and parish bordering the tidal part of the two rivers, together with representatives of user, amenity and conservation groups, in all about twenty-five members. Its aim is to put forward a co-ordinated view to the District and County Councils on matters affecting the estuary. It can be contacted through Hon. Secretary Mr A. W. Wright, Nonsuch, Bridge Lane, Instow, Bideford, Devon EX39 4JD.

Bibliography

General
Devon Coastal Conservation Study (Devon Conservation Forum, 1978)
Devon Wetlands (Devon County Council, 1977)
Coastlines of Devon (Devon County Council, 1980)
Norman, D. and Tucker, V. *Where to watch birds in Devon and Cornwall* (Croom Helm, 1984)
Ovenden, D. and Barrett, J. *Handguide to the sea coast* (Collins, 1981)

Axe estuary
Pulman, G.P.R. *The Book of the Axe*, 1875 (Kingsmead Reprints, facsimile edition 1969)

Otter estuary
Delderfield, E.R. *The Raleigh Country* (Raleigh Press, 1949)
Twelve walks in the Otter Valley (Otter Valley Association, 1983)
The Lower Otter Valley. Sketches on Local History (Otter Valley Association, 1984)
Scutt, W. *A short account of East Budleigh and Hayes Barton: birthplace of Sir Walter Raleigh*, 1936
Sheppard, L. *Raleigh's Birthplace. The story of East Budleigh* (Granary Press, 1983)

Exe estuary
Boalch, G.T. (ed.) *Essays on the Exe Estuary* (Devonshire Association, 1980)
Davies, S. *The Wildlife of the Exe Estuary* (Harbour Books, 1983)
Checklist of the Wildfowl and Waders of the Exe Estuary (Devon Bird Watching and Preservation Society)
Dawlish Warren Nature Reserve (Teignbridge District Council Planning Dept.)
Thomas, P. and Davies, S. *The Exe Estuary – Wildlife in Camera* (Barracuda Books, 1984)

Teign estuary
Chard, J. *Along the Teign* (Bossiney Books, 1981)
Griffiths, G.D. and Griffiths, E.G.C. *History of Teignmouth* (Brunswick Press, 1973)
Teign Estuary Study: discussion document (Teignbridge District Council Planning Dept.)
Trump, H.J. *Westcountry Harbour* (Brunswick Press, 1976)

Dart estuary
Chard, J. *Along the Dart* (Bossiney Books, 1979)
Exploring from Dartmouth – A Walks Booklet (Dartmouth and Kingswear Society)

Freeman, R. *Dartmouth: a New History of the Port and its People* (Harbour Books, 1983)
Hemery, E. *Historic Dart* (David and Charles, 1982)
Soper, T. *Wildlife of the Dart Estuary* (Harbour Books, 1982)

Kingsbridge estuary
Darch, M.D. *The Salcombe and Hope Cove Lifeboat History* (Salcombe and Hope Cove R.N.L.I., 1984)
Fairweather, L. *Salcombe Remembered* (Published by the author, 1980)
Fairweather, L. and Murch, M. *Salcombe Harbour Remembered* (PDS Printers, 1982)
Lubbock, B. *The Last of the Windjammers* (Brown, Son & Ferguson, 1969)
Sixteen Country Walks around Salcombe (Ramblers' Association, 1982)

Avon estuary
Shaw, C.C. *A History of the Parish of Aveton Gifford* (Shaw)

Erme estuary
Risdon, T. *The chronological description or survey of the county of Devon* (Porcupines, new edition 1970)
Britton, J. and Brayley, E. W. *The Beauties of England and Wales. Vol. 4 Devonshire and Dorsetshire*, 1803 (Vernor & Hood)

Tamar estuary
Booker, F. *The Industrial Archaeology of the Tamar Valley* (David & Charles, 1971)
Foot, S. *Following the Tamar* (Bossiney Books, 1980)
Langley, M. and Small, E. *The Port of Plymouth*
Merry, I.D. *Shipping and Trade of the River Tamar* (National Maritime Museum, 1980)

Taw–Torridge estuary
Baxter, J. & Baxter J. *The Bideford, Westward Ho! and Appledore Railway, 1901-1917* (Chard, 1983)
Bear, J. *Appledore, Handmaid of the Sea* (North Devon Museum Trust, 1976)
Elliston-Wright, F.R. *Braunton – a few nature notes* (Barnes, 1932)
Goaman, M. *Old Bideford and District* (Cox and Cox, 1968)
Lamplugh, L. *Barnstaple: town on the Taw* (Phillimore, 1983)
Mayo, R. *The Story of Westward Ho!* (Published by the author, 1983)
Braunton Burrows National Nature Reserve: Flowering Plant List (Nature Conservancy Council, 1977)
Reed, M.A. *Pilton: its past and its people* (D.H. and M.A. Reed, 1977)
Slade, W.J. *Out of Appledore: the autobiography of a coasting shipmaster and shipowner in the last days of wooden sailing ships* (Conway Maritime, 1980)

Tourist Information

Tourist Information Centres throughout Devon provide sources of advice and assistance to help visitors discover and enjoy the county. Centres most relevant to the estuaries are listed below. Most are open only during the summer months, typically April to October. For general information contact Devon Tourism, County Hall, Exeter EX2 4QQ. Tel: Exeter (0392) 53260.

Axe estuary	Seaton Information and Accommodation Centre, The Esplanade, Seaton EX12 2QQ. Tel: Seaton (0297) 21660
Otter estuary	Budleigh Salterton Information Bureau, Rolle Mews Car Park, Fore Street, Budleigh Salterton. Tel: Budleigh Salterton (03954) 5275
Exe estuary	Exmouth Information Centre, Alexandra Terrace, Exmouth EX8 1NZ. Tel: Exmouth (0395) 263744
	Exeter Tourist Information Centre, Civic Centre, Dix's Field, Exeter EX1 1JN. Tel: Exeter (0392) 72434
	Dawlish Tourist Information Centre, The Lawn, Dawlish EX7 9AP. Tel: Dawlish (0626) 863589
Teign estuary	Teignmouth Tourist Information Centre, The Den, Sea Front, Teignmouth TQ14 8BE. Tel: Teignmouth (06267) 6271
	Newton Abbot Tourist Information Centre, 8 Sherborne Road, Newton Abbot. Tel: Newton Abbot (0626) 67494
Dart estuary	Dartmouth and District Tourist Information Centre, Royal Avenue Gardens, Dartmouth. Tel: Dartmouth (08043) 4224 (Postal enquiries to Dartmouth and District Tourist Information Centre, The Dartmouth Enterprise Group, 9 Duke Street, Dartmouth)
	Totnes Tourist Information Caravan, The Plains, Totnes. Tel: Totnes (0803) 863168
Kingsbridge estuary	Kingsbridge Tourist Information Centre, The Quay, Kingsbridge TQ7 1HS. Tel: Kingsbridge (0548) 3195
	Salcombe Information Bureau, Main Road, Salcombe TQ8 8NA. Tel: Salcombe (054 884) 2736
Avon, Erme, Yealm estuaries	Kingsbridge Tourist Information Centre, The Quay, Kingsbridge TQ7 1HS. Tel: Kingsbridge (0548) 3195
	Modbury Tourist Information Centre, 31 Church Street, Modbury PL21 0QR. Tel: Modbury (0548) 830159
Tamar estuary	Plymouth Tourist Information Centre, Civic Centre, Plymouth PL1 2EW. Tel: Plymouth (0752) 264849
	West Devon Tourist Information Centre, Bedford Square, Tavistock PL19 0AE. Tel: Tavistock (0822) 2938
Taw–Torridge estuary	Barnstaple Tourist Information Centre, 20 Holland Square, Barnstaple EX31 1DP. Tel: Barnstaple (0271) 72742
	Bideford Tourist Information Centre, The Quay, Bideford EX39 2EZ. Tel: Bideford (02372) 77676

Follow the Country Code

Enjoy the countryside and respect its life and work
Guard against all risk of fire
Fasten all gates
Keep your dogs under close control
Keep to public paths across farmland
Use gates and stiles to cross fences, hedges and walls
Leave livestock, crops and machinery alone
Take your litter home
Help to keep all water clean
Protect wildlife, plants and trees
Take special care on country roads
Make no unnecessary noise